# Springer Series in Wireless Technology

**Series editor**

Ramjee Prasad, Aarhus University, Herning, Denmark

Springer Series in Wireless Technology explores the cutting edge of mobile telecommunications technologies. The series includes monographs and review volumes as well as textbooks for advanced and graduate students. The books in the series will be of interest also to professionals working in the telecommunications and computing industries. Under the guidance of its editor, Professor Ramjee Prasad of the Center for TeleInFrastruktur (CTIF), Aalborg University, the series will publish books of the highest quality and topical interest in wireless communications.

More information about this series at http://www.springer.com/series/14020

Tamal Chakraborty · Iti Saha Misra
Ramjee Prasad

# VoIP Technology:
# Applications and Challenges

 Springer

Tamal Chakraborty
Future Institute of Engineering and
    Management (FIEM)
Kolkata, India

Iti Saha Misra
Jadavpur University
Kolkata, India

Ramjee Prasad
Future Technologies for Business Ecosystem
    Innovation (FT4BI)
CTIF Global Capsule (CGC), Aarhus
    University
Herning, Denmark

ISSN 2365-4139                   ISSN 2365-4147   (electronic)
Springer Series in Wireless Technology
ISBN 978-3-030-07065-6           ISBN 978-3-319-95594-0   (eBook)
https://doi.org/10.1007/978-3-319-95594-0

This Springer imprint is published by the registered company Springer International Publishing AG
part of Springer Nature
The registered company address is: Gewerbestrasse 11, 6330 Cham, Switzerland

*To my dearest parents, my sister Tanima and brother-in-law Angshuman, my wife Amrita and my niece Antara*

—Tamal Chakraborty

*To my beloved brother, Babun and my spiritual guide, Thakur*

—Iti Saha Misra

*To my grandchildren Ayush, Arya, Akash, Ruchika, and Sneha*

—Ramjee Prasad

# Preface

कर्मण्ये वाधिका रस्ते मा फलेषु कदाचन।

मा कर्म फल हेतु भूर्मा ते सङ्गोऽस्त्व कर्मणि।।

*Karma-naye vadika raste,*

*ma phaleshu kadachana*

*Ma Karma phala he tur bhuh,*

*ma te sangotsva karmanye*

Believe in yourself and do your Karma (action) and success will follow. Doing Karma is in our hands, the result is not. Let not the fruits of action be your motive, nor have attachment to inaction.

—The Bhagvad Gita (2.47)

Since time immemorial, mankind has opened newer frontiers in science and technology leading to innovations that were apparently considered impossible to achieve. Communication is one such realm where revolutionary ideas have been implemented with the sole aim of connecting mankind separated by space and time. At the same time, connectivity—whether the Internet or mobile phones—is increasingly bringing market information, financial aspects, health, and other essential services to remote areas, and is helping to shape people's lives in unprecedented ways. The mobile platform is emerging as the single most powerful way to extend these economic opportunities and key services to millions of people. Globally, these mobile devices have transformed from just a means of voice communication to a multifunctional device that allows users to engage in voice/video conferencing sessions, multimedia entertainment applications, financial services, health tracking, news and information sharing, and many more. As per the International Telecommunication Union (ITU) estimates, the global mobile phone subscriptions reached almost 7 billion in 2014. This infrastructure convergence has lead to the gradual shift of traditional telephony methods toward Voice over IP (VoIP) technology (also known as IP telephony or Internet telephony). VoIP is not only implementable in such integrated devices but also makes international and multibranch voice communication a lot cheaper and easily maintainable.

However, as with any technological advancement, VoIP suffers from scalability issues that are aggravated by its stringent Quality of Service (QoS) requirements. Furthermore, with the advent of emerging networks, VoIP is being implemented over different such platforms that include wireless LAN, Worldwide Interoperability for Microwave Access (WiMAX), cellular networks, Long-Term Evolution (LTE) networks, cognitive radio networks (CRN), 5G networks. Each such system has its own sets of standards and regulations and introduces unique challenges that require a thorough investigation. As the current ICT market is focused toward deploying IoT (Internet-of-things)-enabled solutions in automotive, industrial, educational, and research domains, VoIP is poised to play a prominent role in the design of various such futuristic applications.

Therefore, this book aims to provide a thorough understanding of the VoIP technology with all its merits and challenges in order to highlight its significance in the current context. The readers will not only be conversant with the basic understanding of VoIP applications but will further develop a comprehensive knowledge of the novel research solutions for supporting VoIP services over both conventional and next-generation communication networks.

The several distinguished features of the book are outlined under:

- This book includes detailed discussions on the general architecture of VoIP technology along with its applications and relevance in conventional and emerging wireless communication networks including WLANs, WiMAX, LTE, and CRN.
- The fundamental and theoretical concepts on the call signaling protocols which include SIP, H.323, and MEGACO coupled with an extensive comparative analysis among these protocols to facilitate a better understanding of the intricacies involved in VoIP communication are discussed.
- It focuses on the voice quality issues, in particular, the Quality of Service (QoS) characteristics including the QoS framework, policies, principles, and the VoIP QoS metrics such as delay, jitter, packet loss, R-factor, and Mean Opinion Score (MOS).
- It provides test-bed design and implementation of VoIP telephony over wireless LANs spanning multiple buildings with both real-time and offline monitoring solutions to serve as the platform for comparative performance evaluation and validation of the proposed research formulations.
- Performance test studies of VoIP communication are given in terms of the QoS metrics under the developed test-bed platform with respect to the standard call signaling protocols as well as under different codec parameters and buffer sizes.
- QoS management for VoIP calls under dynamic and congested network scenarios is described through simulation and test-bed execution of novel research solutions ranging from the parameterization of access points, soft phones, and

active queues to codecs and heuristic search based optimization tools along with adaptive packet size variations and adaptive jitter buffer algorithms.

- Special emphasis is outlined on the upcoming wireless communication networks, specifically the next-generation cognitive radio networks (CRN) with detailed discussions on the challenges and research inclinations for implementing VoIP services over this CRN platform substantiated by the innovative design of a model network for VoIP-based CR system and analysis of significant parameters concerning both VoIP QoS metrics and CR parameters.

The book is organized as follows:

Chapter 1 presents an overview of the Voice over IP technology and elaborately describes its relevance in the modern society. The motivation for the selection of VoIP as a communication service is also established. Special emphasis has been laid on deploying VoIP over different networks including WLANs, WiMAX, LTE, CRN. Finally, the evolution of VoIP and its penetration in public and private sectors have been tracked over the years that justify the research initiatives pursued in the subsequent chapters.

Chapter 2 describes the fundamental call signaling protocols used for call management in VoIP communication including H.323, SIP, and MEGACO. These protocols are also evaluated for performance efficiency with respect to various aspects such as call setup and management issues, addressing and security aspects, complexity analysis. Finally, the current status of these protocols after various ratifications and updates is presented.

Chapter 3 deals with one of the most important aspects in VoIP communication: Quality of Service (QoS) maintenance. In this regard, the various QoS categories, metrics, and policies are defined and discussed with respect to VoIP applications. QoS implementation policies are introduced based on a QoS framework. The fundamental difference with QoE is also highlighted. Finally, the current approach to problem-solving in regard to QoS management is demonstrated.

Chapter 4 provides an overview of wireless LANs and discusses the pros and cons of implementing VoIP applications over these networks. Precisely, it highlights the current design challenges and shortcomings in WLAN that require careful addressing for a successful hosting of VoIP calls. This is also substantiated by a detailed description of VoIP-based WLAN test bed that is designed exclusively for laboratory experiments. A real-time test bed-based performance analysis is carried out to study some of the most important factors that influence and tune the performance of the VoIP soft phones in WLAN.

The performance of wireless LANs is greatly affected by path-loss, RF interference, and other sources of signal attenuation, in addition to network congestion. The primary factors involved in effective real-time communication, namely delay and loss, must be within certain controlled limits in such a scenario.

To the end, Chap. 5 analyzes the various factors driving IEEE 802.11b access points through extensive simulations and thereafter develops optimization techniques to configure the parameters of the access points. The configured parameters are then implemented in the SIP-enabled test bed to provide optimum Voice over IP

(VoIP) performance. Thereafter, the parameters of VoIP soft phones are analyzed and optimized to support both voice and video calls. Finally, active queue management system is applied for ensuring sufficient QoS of VoIP calls. To this end, an optimization algorithm is proposed for proper selection of threshold queuing parameters and successfully implemented in the hardware test bed.

As optimizations in the field of QoS maintenance continue to evolve and mitigate the effects of unpredictable nature of networks, implementing them in proper sequence to achieve the highest performance efficiency is the biggest challenge and is henceforth addressed in Chap. 6. Moreover, each such QoS implementation mechanism must be maintained adaptively to cope up with variations in the network or changes in the user scenarios. To resolve conflicts, the system must capture system policies, including end-to-end scheduling policies, policies to decide which application's QoS to degrade when there are not enough resources to provide the desired QoS to all applications, and admission control policies. In this chapter, an optimization algorithm is proposed driven by real-time heuristic incremental state-space search that fulfills the aim of maintaining adaptive QoS in multiple call scenarios and under diverse network conditions by applying the available QoS optimization techniques in proper sequence.

Codec is an essential component in VoIP technology, and therefore, the dream of achieving perfect call quality in VoIP cannot be fulfilled without addressing the issues related to codecs. Accordingly, Chap. 7 focuses on codec parameters and evaluates their impact on VoIP QoS. In multirate WLANs, users can suffer transmission rate changes due to the link adaptation mechanism. This results in a variable capacity channel, which is unsuitable for hosting VoIP calls and can cause serious quality of service (QoS) degradation in all the active calls. Various codec adaptation mechanisms have been proposed as a solution to this problem, as well as to solve the congestion problems on WLAN environments. However, most of them follow a reactive strategy based on packet loss information to switch codecs. In this chapter, a special codec parameter, namely 'pps', has been analyzed and an algorithm based on proactive strategy is proposed to decrease the pps in the WLAN access points so as to reduce the packet loss due to buffer overflow.

Chapter 8 lays special emphasis on jitter which has a significant influence on the overall voice call quality. Variation in network characteristics introduces jitter to the propagating voice packets. Jitter hampers voice quality and makes the VoIP call uncomfortable to the user. Often buffers are used to store the received packets for a short time before playing them at equal spaced intervals to minimize jitter. Choosing optimum buffering time is essential for reducing the added end-to-end delay and number of discarded packets. In this chapter, some established adaptive jitter buffer play-out algorithms have been studied and a new algorithm has been proposed to address the shortcomings of the established ones. Further studies have been conducted for finding the optimum sliding window size for the proposed algorithm. The designed algorithm kept jitter within tolerable limit along with a significant reduction of delay and loss as compared to existing algorithms in the literature.

Chapter 9 has explored the use of variable packet payload sizes as an alternative to the already existing methods to sustain VoIP calls under congested network scenarios. OPNET Modeler 14.5.A has been used for carrying out extensive analysis for voice quality in terms of MOS and end-to-end delays. An algorithm has been proposed that utilizes the RTCP receiver reports to assess the network conditions and select the optimum packet payload size. The proposed algorithm has been implemented in OPNET simulation model, and exhaustive simulations have been undertaken to test its efficiency. The algorithm has provided satisfactory results while maintaining its transparency with respect to the end user.

Finally, Chap. 10 has explored the novel paradigm of deploying VoIP over the next-generation networks, in particular, the cognitive radio networks. The concept of opportunistic communication for exploiting the spectrum underutilization problem through CRN is discussed, and its applications are studied. To this end, the motivation for incorporating VoIP services over CRN is discussed and the inherent challenges introduced by such integration are studied in detail. Finally, this chapter provides a basic overview of the design principle for implementing VoIP applications in CRN and applies this principle toward developing the simulation model in OPNET. This model helps in analyzing the VoIP QoS metrics in conjunction with the CRN parameters and further paves the way toward conducting exhaustive research activities in future.

Kolkata, India                                                                          Tamal Chakraborty
Kolkata, India                                                                               Iti Saha Misra
Herning, Denmark                                                                        Ramjee Prasad

# Acknowledgements

Writing a book is an all-encompassing as well as a time-consuming task which certainly could not have been achieved without the support and encouragement from families and friends. Special thanks to Tamal's beloved family especially his parents Mr. Tapas Kumar Chakraborty and Mrs. Pratima Chakraborty for being the pillar of strength and inspiration, to the lovely family of Prof. Iti Saha Misra, for their understanding and motivation and Vaishnavi Inamdar for her support in refining the book.

The authors would like to express sincere gratitude and respect to Prof. Dr. S. K. Sanyal, Dept. of ETCE, Jadavpur University, Kolkata, whose truly scientific intuition and meaningful discussions during the laboratory experiments on VoIP research led to the realization of Chaps. 5–7 and Mr. Atri Mukhopadhyay for his valuable input in Chaps. 8 and 9.

Our thanks to Pulak Roy, Sayan Sengupta, Aritra Chatterjee, Bodhisatwa Biswas, Subhojit Mukherjee, Arnob Maity, Dr. Budhaditya Bhattacharyya, Tanumay Manna, Anindita Kundu, Tanmay Kundu, Sudipta Dey, Arijeet Ghosh, Sreya Ghosh for their help and inquisitiveness that kept us active and up to date with the recent technologies.

# Contents

**1  Overview of VoIP Technology** ..................................... 1
   1.1    Introduction ........................................... 1
   1.2    What Is Voice Over IP (VoIP)? .......................... 2
   1.3    VoIP: Why Implement It? ............................... 4
   1.4    Evolution of VoIP..................................... 5
   1.5    How VoIP Works? ..................................... 7
   1.6    VoIP Applications..................................... 12
   1.7    VoIP: Present and Future................................ 14
         1.7.1   VoIP Over WLAN ........................... 15
         1.7.2   VoIP Over WiMAX .......................... 16
         1.7.3   Voice Over LTE............................. 18
         1.7.4   VoIP Over Cognitive Radio Network.............. 18
   1.8    VoIP Popularity ....................................... 20
   1.9    Summary ............................................ 23
   References ................................................. 23

**2  VoIP Protocol Fundamentals** ................................... 25
   2.1    Introduction ........................................... 25
   2.2    H.323.............................................. 26
         2.2.1   Elements ................................. 27
         2.2.2   Protocol Suite............................. 27
         2.2.3   Call Flow................................ 29
         2.2.4   Enhancements ............................ 32
   2.3    Session Initiation Protocol .............................. 33
         2.3.1   SIP Actors ............................... 35
         2.3.2   SIP Structure ............................. 35
         2.3.3   SIP Message Type ......................... 36
         2.3.4   SIP Call Flows............................ 37

2.4      Megaco (H.248) . . . . . . . . . . . . . . . . . . . . . . . . .        39
         2.4.1    Call Flow Description . . . . . . . . . . . . . . . . . . . .   39
         2.4.2    Command Format . . . . . . . . . . . . . . . . . . . . . .     41
         2.4.3    Megaco and PSTN . . . . . . . . . . . . . . . . . . . .        41
2.5      Security Issues in H.323 and SIP . . . . . . . . . . . . . . . . .      42
2.6      H.323 and SIP: Comparison . . . . . . . . . . . . . . . . . . .         43
         2.6.1    Addressing . . . . . . . . . . . . . . . . . . . . . . . . .   43
         2.6.2    Complexity . . . . . . . . . . . . . . . . . . . . . . . .     44
         2.6.3    Call Setup . . . . . . . . . . . . . . . . . . . . . . . .     44
         2.6.4    Extensibility . . . . . . . . . . . . . . . . . . . . . . . . 44
2.7      Current Status of H.323 and SIP . . . . . . . . . . . . . . . . .       45
         2.7.1    H.323 . . . . . . . . . . . . . . . . . . . . . . . . . . . .  45
         2.7.2    SIP . . . . . . . . . . . . . . . . . . . . . . . . . . . . .  45
2.8      Summary . . . . . . . . . . . . . . . . . . . . . . . . . . . .         46
References . . . . . . . . . . . . . . . . . . . . . . . . . . . . . . . . . .   46

3   **Quality of Service Management—Design Issues** . . . . . . . . . . . . .    49
3.1      Introduction . . . . . . . . . . . . . . . . . . . . . . . . . . .      49
3.2      QoS as Service . . . . . . . . . . . . . . . . . . . . . . . . . .      50
3.3      QoS Framework . . . . . . . . . . . . . . . . . . . . . . . . .         52
3.4      QoS Parameters for VoIP Applications . . . . . . . . . . . . . . .      53
3.5      QoS Implementation Policies . . . . . . . . . . . . . . . . . . . .     60
3.6      Quality of Experience . . . . . . . . . . . . . . . . . . . . . . .     65
3.7      Current Approach . . . . . . . . . . . . . . . . . . . . . . . . .      66
3.8      Summary . . . . . . . . . . . . . . . . . . . . . . . . . . . .         68
References . . . . . . . . . . . . . . . . . . . . . . . . . . . . . . . . . .   68

4   **VoIP Over Wireless LANs—Prospects and Challenges** . . . . . . . . . .     71
4.1      Introduction . . . . . . . . . . . . . . . . . . . . . . . . . . .      71
4.2      Overview of Wireless LANs . . . . . . . . . . . . . . . . . . . .       71
4.3      Important Aspects Regarding Voice Over WLAN . . . . . . . . . .         73
4.4      Design Challenges for VoIP Services in WLAN . . . . . . . . . . .       75
         4.4.1    VoIP Over Evolving WLAN Standards . . . . . . . . . . .        75
         4.4.2    System Capacity in WLAN . . . . . . . . . . . . . . . . .      75
         4.4.3    Inadequacy of PCF . . . . . . . . . . . . . . . . . . . . .    76
         4.4.4    Admission Control Issues . . . . . . . . . . . . . . . . . .   77
         4.4.5    Security Challenges . . . . . . . . . . . . . . . . . . . . .  77
         4.4.6    Sleep Mode for WLAN and Power-Saving Issues . . . .            77
         4.4.7    Handoff Problem in WLAN . . . . . . . . . . . . . . . . .      78
4.5      Prospective Solutions for VoIP Policy Design in WLAN . . . . .          78
4.6      Related Works . . . . . . . . . . . . . . . . . . . . . . . . . . .     79
4.7      Test-Bed Model for VoIP Deployment Over WLAN . . . . . . . .            80
         4.7.1    Hardware Modules . . . . . . . . . . . . . . . . . . . . .     81
         4.7.2    Software Elements . . . . . . . . . . . . . . . . . . . . .    84

|       | 4.8   | Performance Analysis in the Test Bed | 85 |
|       |       | 4.8.1 Analysis of Call Signaling Protocols | 85 |
|       |       | 4.8.2 Analysis of Codec Parameters and Buffer Size | 87 |
|       | 4.9   | Summary | 90 |
|       | References | | 90 |

**5  Technique for Improving VoIP Performance Over Wireless LANs** ......... 95

|   | 5.1 | Introduction | 95 |
|   | 5.2 | Proposed Algorithm for Access Points | 96 |
|   |     | 5.2.1 Analysis | 96 |
|   |     | 5.2.2 Proposed Optimization Technique | 99 |
|   |     | 5.2.3 Implementation | 103 |
|   | 5.3 | Proposed Technique for Node Parameters | 107 |
|   |     | 5.3.1 Analysis | 108 |
|   |     | 5.3.2 Optimization Technique | 109 |
|   |     | 5.3.3 Implementation | 113 |
|   | 5.4 | Proposed Active Queue Management Policy | 115 |
|   | 5.5 | Summary | 120 |
|   | References | | 121 |

**6  Optimizing VoIP in WLANs Using State-Space Search** ......... 123

|   | 6.1 | Introduction | 123 |
|   | 6.2 | Motivation | 124 |
|   | 6.3 | Why Use State-Space Approach? | 126 |
|   | 6.4 | Optimization of VoIP Call Using Dynamic Search | 126 |
|   |     | 6.4.1 Proposed Technique | 127 |
|   |     | 6.4.2 Implementation of the Algorithm | 130 |
|   | 6.5 | Implementation of Learning Strategy for QoS Enhancement of VoIP Calls | 140 |
|   |     | 6.5.1 Proposed Algorithm | 141 |
|   |     | 6.5.2 Implementation of the Algorithm | 143 |
|   | 6.6 | Benefits of the Proposed Algorithm | 144 |
|   | 6.7 | Summary | 145 |
|   | References | | 146 |

**7  Optimization of Codec Parameters to Reduce Packet Loss Over WLAN** ......... 149

|   | 7.1 | Introduction | 149 |
|   | 7.2 | VoIP Codecs | 150 |
|   |     | 7.2.1 Overview | 150 |
|   |     | 7.2.2 Codec Parameters | 150 |
|   |     | 7.2.3 Evaluation of Codec | 151 |
|   |     | 7.2.4 Codec Compression Techniques | 154 |

7.3    Motivation ........................................  159
7.4    Analysis ..........................................  159
7.5    Algorithm .........................................  162
       7.5.1    Assumptions ...............................  162
       7.5.2    Algorithm .................................  162
7.6    Implementation of the Algorithm .....................  164
7.7    Summary ..........................................  166
References ..............................................  167

**8    QoS Enhancement Using an Adaptive Jitter Buffer Algorithm
       with Variable Window Size ...............................  169**
8.1    Introduction ......................................  169
8.2    Background Study ...................................  171
8.3    Related Work ......................................  173
       8.3.1    Exponential Average Algorithm (EXP-AVG) .......  173
       8.3.2    Fast Exponential Average Algorithm
                (F-EXP-AVG) ...............................  174
       8.3.3    Minimum Delay Algorithm (Min-D) .............  174
       8.3.4    Spike Detection Algorithm (Spike-Det) ...........  174
       8.3.5    Window Algorithm ..........................  176
8.4    The Simulation Setup ...............................  177
8.5    Analysis of the Existing Adaptive Jitter Buffer Playout
       Algorithms ........................................  178
8.6    Proposed Adaptive Jitter Buffer Playout Algorithm .........  180
8.7    Results ...........................................  183
       8.7.1    Implementing the Algorithm with a Window
                Size of 100 ...............................  183
8.8    Effect of Various Window Sizes .......................  185
8.9    Comparative Results with the Other Analyzed Algorithms ....  188
8.10   Summary ..........................................  189
References ..............................................  190

**9    Adaptive Packetization Algorithm to Support VoIP Over
       Congested WLANS .....................................  193**
9.1    Introduction ......................................  193
9.2    Background Literature ...............................  194
9.3    The Simulator .....................................  196
       9.3.1    Simulation Setup ..........................  197
       9.3.2    Simulation Scenarios .......................  199
9.4    Proposed Packetization Algorithm .....................  206
9.5    Implementation and Results ...........................  213

|  |  | 9.5.1 | Implementation | 213 |
|  |  | 9.5.2 | Results | 213 |
|  | 9.6 | Summary |  | 216 |
|  | References |  |  | 216 |

**10  VoIP Over Emerging Networks: Case Study with Cognitive Radio Networks** ............................................ 217

|  | 10.1 | Introduction | 217 |
|  | 10.2 | What Is Cognitive Radio Network? | 218 |
|  | 10.3 | VoIP Over CRN: Why?? | 221 |
|  | 10.4 | Challenges Toward Deploying VoIP Over CRN | 224 |
|  | 10.5 | Key Focus Areas | 224 |
|  | 10.6 | System Parameters in CRN | 226 |
|  |  | 10.6.1 Spectral Efficiency | 226 |
|  |  | 10.6.2 System Capacity | 226 |
|  |  | 10.6.3 Energy Efficiency | 227 |
|  |  | 10.6.4 CR Timing Cycle | 227 |
|  |  | 10.6.5 Spectrum Handoff Delay | 228 |
|  | 10.7 | Design Principle: An Overview | 228 |
|  | 10.8 | Model Overview for VoIP Deployment Over CRN | 230 |
|  |  | 10.8.1 Simulation Setup | 232 |
|  |  | 10.8.2 Discussion of Simulation Results | 234 |
|  | 10.9 | Applications | 235 |
|  | 10.10 | Summary | 237 |
|  | References |  | 237 |

# About the Authors

**Dr. Tamal Chakraborty** is currently Assistant Professor in the Department of Computer Science and Engineering, Future Institute of Engineering and Management (FIEM), Garia, Kolkata. Prior to this, he was Assistant Professor of BITS Pilani Campus. He has received his Ph.D. (Engg.) as DST INSPIRE Fellow from the Department of Electronics and Telecommunication Engineering, Jadavpur University, India. He was awarded the University Medal for his M.Tech. from Jadavpur University and has received his B.Tech. in Computer Science and Engineering from the West Bengal University of Technology, Kolkata. He is the recipient of the IEEE Young Scientist Award (first prize) 2015 and also the Best Paper Award in ACM International Conference 2012. He has also received the Best Volunteer Award 2017 from IEEE ComSoc Society, Kolkata Chapter, India as well as RIG award, 2017 from BITS Pilani, India. He has published several SCI and SCI-E indexed journal papers (IEEE, Elsevier, Academy Publishers), 16 conference proceeding papers (IEEE, Springer, ACM), and two chapters (Springer). He has attended and delivered several presentations at International Conferences (India, China, Mauritius, Bangladesh), Forums, and Invited Talks on upcoming research prospects and applications. His research interests include wireless communication and networking, Voice over IP, Quality of Service studies, cognitive radio networks, test-bed prototype design and implementation, intelligent transportation systems, pervasive computing, and related emerging areas in computing and communication, where he has filed several patents. He is IEEE

Member and has served as Reviewer of several SCI indexed journals (including the IEEE Transactions on Wireless Communication, Journal on Selected Areas in Communication, IEEE Systems Journal, Computer and Electrical Engineering Journal Elsevier). As part of his cocurricular activities, he is a literary enthusiast with publications in International Literary Magazines and also an active quizzard holding the second rank in the National Quiz Competition held by Petroleum Conservation Research Association, Government of India.

**Dr. Iti Saha Misra** is presently holding the post of Professor in the Department of Electronics and Telecommunication Engineering, Jadavpur University, Kolkata, India. She was the immediate past Head of the same department. She completed her B.Tech. in Radio Physics and Electronics from Calcutta University (1989) and Masters in Telecommunication Engineering from Jadavpur University (1991), Kolkata, India. After the completion of Ph.D. in Engineering in the field of Microstrip Antennas from Jadavpur University (1997), she is actively engaged in teaching since 1997. She has 25+ years of research experience in the field of wireless/mobile communication, microwaves, microstrip antennas, numerical computational methods, and computer networks. She has developed the Broadband Wireless Communication Laboratory in the Department of Electronics and Telecommunication Engineering of Jadavpur University to carry out advanced research work in wireless communication domain. She is Coordinator of Communication Specialization in the Department of ETCE for postgraduate course.

She is the recipient of prestigious Career award for Young teachers by All India Council for Technical Education (AICTE) in 2004, obtained IETE Gowri memorial award in 2007 in the best paper category for the topic of "4G networks: Migration to the Future", co-author of several best paper awards in the wireless communication domain. She has supervised 20 doctoral research students (nine awarded, two under submission, two under completion, and seven ongoing) and 50 theses at master's level in the field of mobile communication and antennas. Under her supervision, two Ph.D. students received the Young Scientist Award from IEEE URSI International Radio Conference out of their research presentation in the years 2014 and 2015, consecutively.

Her current research interests are in the areas of cognitive radio networks, VoIP and video communication over cognitive platform, wireless body area networks (WBAN), low-cost and energy-efficient IoT solutions, call admission control and packet scheduling for wireless and BWA networks, radio resource management for 5G networks, channel allocation in macro/femto networks, mobility management, network architecture and protocols, integration architecture of WLAN and 3G networks, etc. Her other research activities are related to design optimization of wire antennas using numerical techniques. She has authored more than 210 research papers in refereed journals and international conferences and has filed several patents. She is the author of a widely acclaimed textbook on "Wireless Communication and Networks: 3G and Beyond" published by McGraw Hill. She has supervised several projects from Government of India, such as DST FIST, DST PURSE, UGC-UPE, AICTE.

She has visited many countries like USA, UK, France, Australia, Singapore, Prague, Malaysia, Thailand, Indonesia, and Bangladesh to present research papers in the IEEE and other reputed international conferences, and delivered several invited lectures at international platform. She has visited NTRC Laboratory at Nanyang Technological University, Singapore, in January 2006 for one month under TEQIP of Jadavpur University.

She is Senior Member of IEEE, Present ComSoc Chairperson, Kolkata Chapter, the founder Chair of Women in Engineering Affinity Group, IEEE Kolkata Section. Under her leadership, WIE Affinity Group won the first prize in 2007 and IEEE ComSoc Kolkata Chapter won the 2015 and 2017 Chapter Achievement Award. She has served as the Reviewer of several SCI indexed journals such as IEEE Transactions on Wireless Communication, IEEE Transaction of Antenna and Propagation, Journal of Wireless Networks, Springer, European Transaction on Telecommunications, Wiley, IEEE Journal APWL, International Journal of RF and Microwave Computer-Aided Engineering along with several IEEE International conference paper reviews like Globecom, VTC, ICON.

**Dr. Ramjee Prasad** Fellow IEEE, IET, IETE, and WWRF, is Professor of Future Technologies for Business Ecosystem Innovation (FT4BI) in the Department of Business Development and Technology, Aarhus University, Herning, Denmark. He is Founder President of the CTIF Global Capsule (CGC). He is also Founder Chairman of the Global ICT Standardisation Forum for India, established in 2009. GISFI has the purpose of increasing of the collaboration between European, Indian, Japanese, North American, and other worldwide standardization activities in the area of information and communication technology (ICT) and related application areas.

He has been honored by the University of Rome "Tor Vergata", Italy, as a Distinguished Professor of the Department of Clinical Sciences and Translational Medicine on March 15, 2016. He is Honorary Professor of University of Cape Town, South Africa, and University of KwaZulu-Natal, South Africa.

He has received Ridderkorset af Dannebrogordenen (Knight of the Dannebrog) in 2010 from the Danish Queen for the internationalization of top-class telecommunication research and education.

He has received several international awards such as IEEE Communications Society Wireless Communications Technical Committee Recognition Award in 2003 for making contribution in the field of "Personal, Wireless and Mobile Systems and Networks", Telenor's Research Award in 2005 for impressive merits, both academic and organizational within the field of wireless and personal communication, 2014 IEEE AESS Outstanding Organizational Leadership Award for "Organizational Leadership in developing and globalizing the CTIF (Center for TeleInFrastruktur) Research Network".

He has been Project Coordinator of several EC projects, namely MAGNET, MAGNET Beyond, eWALL.

He has published more than 40 books, 1000+ journal and conference publications, more than 15 patents, over 100 Ph.D. graduates and larger number of masters (over 250). Several of his students are today worldwide telecommunication leaders themselves.

# Abbreviations

| | |
|---|---|
| A/D | Analog-to-digital |
| ADPCM | Adaptive differential pulse code modulation |
| AM | Amplitude modulation |
| AP | Access point |
| ARJ | Admission Reject |
| ARQ | Admission Request |
| ASCII | American Standard Code for Information Interchange |
| ASN.1 | Abstract Syntax Notation.1 |
| ATA | Analog terminal adapter |
| B2BUA | Back-To-Back User Agent |
| BER | Basic encoding rules |
| BFS | Breadth-first search |
| BNF | Binary message form |
| BS | Base station |
| BSS | Basic service set |
| BV32 | BroadVoice 32 |
| CAC | Call admission control |
| CBR | Constant bit rate |
| CELP | Code-excited linear prediction compression |
| CFP | Contention free period |
| CNAME | Canonical name |
| CoS | Class of Service |
| CP | Contention period |
| CPE | Consumer premise equipment |
| CQ | Custom queuing |
| CRN | Cognitive radio network |
| cRTP | Compressed RTP |
| CS-ACELP | Conjugate-structure algebraic-code-excited linear prediction |
| CSFB | Circuit-switched fall back |
| CSMA/CA | Carrier sense multiple access with collision avoidance |

| | |
|---|---|
| CTI | Computer telephony integration |
| CTS | Clear to send |
| DCF | Distributed coordination function |
| DFS | Depth-first search |
| DiffServ | Differentiated service |
| DRR | Deficit round robin |
| DS | Distribution system |
| DSCP | Differentiated services code point |
| DSL | Dynamic spectrum leasing |
| DSP | Digital signal processor |
| DSSS | Direct-sequence spread spectrum |
| EIRP | Equivalent isotropically radiated power |
| EPC | Evolved packet core |
| ESS | Extended service set |
| ETSI | European Telecommunications Standards Institute |
| eUTRAN | Evolved UMTS terrestrial radio access network |
| EXP-AVG | Exponential average algorithm |
| FCC | Federal Communications Commission |
| FEC | Forward error correction |
| F-EXP-AVG | Fast exponential average algorithm |
| FHSS | Frequency-hopping spread spectrum |
| FIFO | First in, first out |
| FM | Frequency modulation |
| GSM | Global System for Mobile Communications |
| HTML | Hypertext markup language |
| HTTP | Hypertext transfer protocol |
| IBSS | Independent basic service set |
| ICMP | Internet control message protocol |
| IDC | Index for dispersion of counts |
| IETF | Internet Engineering Task Force |
| IM | Internet messenger |
| IMS | IP multimedia system |
| IntServ | Integrated service |
| IP | Internet protocol |
| IPSec | IP security protocol |
| IR | Infrared |
| ISM | Industrial, Scientific, and Medical Band |
| ITU-T | International Telecommunication Union, the Telecommunication division |
| LAN | Local area network |
| LPA* | Lifelong Planning A* |
| LPC | Linear predictive coding |
| LRTA* | Learning Real-Time A* |
| LTE | Long-Term Evolution |
| MAC | Media Access Control |

| | |
|---|---|
| MCU | Multipoint Control Unit |
| MG | Media gateway |
| MGC | Media gateway controller |
| Min-D | Minimum delay algorithm |
| MLPP | Multilevel precedence and preemption |
| MMPP | Markov-modulated Poisson process |
| MMS | Multimedia messaging service |
| MMUSIC | Multiparty multimedia session control |
| MOS | Mean Opinion Score |
| MP-MLQ | Multipulse multilevel quantization |
| MSDU | MAC service data unit |
| MTU | Maximum transmission unit |
| MWI | Message waiting indicator |
| NAT | Network address translation |
| NGN | Next-generation network |
| OPNET | Optimized Network Engineering Tool |
| PBX | Private branch exchange |
| PCF | Point coordination function |
| PCM | Pulse code modulation |
| PCT | Private communication technology |
| PDU | Protocol data unit |
| PER | Packet encoding rules |
| PGP | Pretty Good Privacy |
| PHB | Per-hop behavior |
| PHY | Physical |
| PLC | Packet loss concealment |
| PQ | Priority queuing |
| PSTN | Public switched telephone network |
| PU | Primary user |
| Q.O. | Queue occupancy |
| QoE | Quality of Experience |
| QoP | Quality of Perception |
| QoS | Quality of Service |
| RAS | Registration, admissions, and status |
| RED | Random early detection |
| RF | Radio frequency |
| ROCCO | RObust Checksum-based header COmpression |
| ROHC | RObust Header Compression |
| RPE-LTP | Regular-Pulse Excitation Long-Term Predictor |
| RR | Receiver report |
| RRS | Recursive random search |
| RSVP | Resource reservation protocol |
| RTCP | Real-time control protocol |
| RTP | Real-time transport protocol |
| rtPS | Real-time polling service |

| | |
|---|---|
| RTS | Request to send |
| SCN | Switched circuit network |
| SIP | Session initiation protocol |
| SMS | Short message service |
| SP | Single space |
| Spike-Det | Spike detection algorithm |
| SR | Sender report |
| SRTP | SIP-signaled secure RTP |
| SSL | Secure socket layer |
| SU | Secondary user |
| SV-LTE | Simultaneous Voice over LTE |
| TCP/IP | Transmission-control protocol/Internet protocol |
| TDM | Time-division multiplexing |
| TLS | Transport layer security |
| ToIP | Text-over-IP |
| ToS | Type of Service |
| TSAP | Transport services access point |
| TXOP | Transmission opportunity |
| UAC | User agent client |
| UAS | User agent server |
| UDP | User datagram protocol |
| UGS | Unsolicited grant service |
| UHF | Ultra-high frequency |
| URI | Uniform resource identifier |
| VBR | Variable bit rate |
| VHF | Very high frequency |
| VLAN | Virtual local area network |
| VoIP | Voice over IP |
| VoLGA | Voice over LTE via GAN |
| VoLTE | Voice over LTE |
| VSC | Virtual service communities |
| WBAN | Wireless body area network |
| WFQ | Weighted fair queuing |
| WiMAX | Worldwide Interoperability for Microwave Access |
| WLAN | Wireless local area network |
| WRAN | Wireless regional area network |
| WWW | World Wide Web |

# List of Figures

Fig. 1.1     Communication aspects for different applications . . . . . . . . . . .    3
Fig. 1.2     Evolution of VoIP technology . . . . . . . . . . . . . . . . . . . . . . . .    7
Fig. 1.3     Working mechanism behind VoIP communication . . . . . . . . . .    8
Fig. 1.4     Complete VoIP communication between the sender
and the receiver . . . . . . . . . . . . . . . . . . . . . . . . . . . . . . . . . .    9
Fig. 1.5     Development of VoIP packet . . . . . . . . . . . . . . . . . . . . . . . . . .   10
Fig. 1.6     Fundamental elements of a VoIP Network . . . . . . . . . . . . . . .   11
Fig. 1.7     Role of VoIP in multiconferencing system . . . . . . . . . . . . . . .   13
Fig. 1.8     Application areas for VoIP . . . . . . . . . . . . . . . . . . . . . . . . . . .   15
Fig. 1.9     Steps for deploying VoIP over WLAN . . . . . . . . . . . . . . . . . . .   16
Fig. 1.10    Architectural overview of VoIP in cognitive radio
network . . . . . . . . . . . . . . . . . . . . . . . . . . . . . . . . . . . . . . . . .   19
Fig. 1.11    Percentage utilization of Internet by users in different
countries . . . . . . . . . . . . . . . . . . . . . . . . . . . . . . . . . . . . . . . .   20
Fig. 1.12    Mobile telephony and mobile broadband users in developing
and developed countries . . . . . . . . . . . . . . . . . . . . . . . . . . . . .   21
Fig. 1.13    Volumes of VoIP traffic generated over the years . . . . . . . . . .   21
Fig. 1.14    Total number of VoIP subscribers for various quarters
of 2011–2013 . . . . . . . . . . . . . . . . . . . . . . . . . . . . . . . . . . . .   22
Fig. 1.15    Distribution of VoIP users across continents . . . . . . . . . . . . . .   22
Fig. 2.1     H.323 elements and their interactions . . . . . . . . . . . . . . . . . . .   28
Fig. 2.2     H.323 protocol layers . . . . . . . . . . . . . . . . . . . . . . . . . . . . . .   28
Fig. 2.3     H.323 direct call flow model . . . . . . . . . . . . . . . . . . . . . . . . .   30
Fig. 2.4     H.323 gatekeeper-routed call flow model . . . . . . . . . . . . . . . .   31
Fig. 2.5     SIP functionalities . . . . . . . . . . . . . . . . . . . . . . . . . . . . . . . .   34
Fig. 2.6     SIP call proxying . . . . . . . . . . . . . . . . . . . . . . . . . . . . . . . . .   37
Fig. 2.7     SIP call redirection . . . . . . . . . . . . . . . . . . . . . . . . . . . . . . . .   38
Fig. 2.8     Megaco entities . . . . . . . . . . . . . . . . . . . . . . . . . . . . . . . . . . .   39
Fig. 2.9     Operation of Megaco in PSTN via IP networks . . . . . . . . . . . .   42
Fig. 3.1     Traffic conditioners for DiffServ QoS policy . . . . . . . . . . . . . .   51
Fig. 3.2     Logical flow for IntServ QoS policy . . . . . . . . . . . . . . . . . . . .   52

Fig. 3.3      QoS framework. . . . . . . . . . . . . . . . . . . . . . . . . . . . . . . . . . . . .    54
Fig. 3.4      Delay components in VoIP system . . . . . . . . . . . . . . . . . . . . . .    56
Fig. 3.5      Compressed RTP mechanism . . . . . . . . . . . . . . . . . . . . . . . . . .    61
Fig. 3.6      Traffic policing and traffic shaping . . . . . . . . . . . . . . . . . . . . . .    62
Fig. 3.7      Fragmentation and interleaving . . . . . . . . . . . . . . . . . . . . . . . . .    63
Fig. 3.8      Playout buffer . . . . . . . . . . . . . . . . . . . . . . . . . . . . . . . . . . . . .    64
Fig. 3.9      Definition domain for quality of experience (QoE). . . . . . . . . .    66
Fig. 3.10     SWAN basic architecture . . . . . . . . . . . . . . . . . . . . . . . . . . . . .    67
Fig. 4.1      Architecture of IEEE 802.11 network . . . . . . . . . . . . . . . . . . . .    72
Fig. 4.2      Relation between WLAN system capacity and VoIP QoS . . . .    76
Fig. 4.3      Domain diagram . . . . . . . . . . . . . . . . . . . . . . . . . . . . . . . . . . . .    81
Fig. 4.4      Variation in packet loss for every access point . . . . . . . . . . . . .    86
Fig. 4.5      Variation in MOS for every access point . . . . . . . . . . . . . . . . .    87
Fig. 4.6      Variation in $R$-factor for every access point . . . . . . . . . . . . . . .    87
Fig. 4.7      Variation in delay, loss, and MOS with time for high
              bit rate codec . . . . . . . . . . . . . . . . . . . . . . . . . . . . . . . . . . . . . .    88
Fig. 4.8      Variation in delay, loss, and MOS with time for low
              bit rate codec . . . . . . . . . . . . . . . . . . . . . . . . . . . . . . . . . . . . . .    88
Fig. 4.9      Variation in delay, loss, and MOS with time for adaptive
              bit rate codec . . . . . . . . . . . . . . . . . . . . . . . . . . . . . . . . . . . . . .    89
Fig. 4.10     Variation in packet loss with decrease in buffer size
              over time. . . . . . . . . . . . . . . . . . . . . . . . . . . . . . . . . . . . . . . . . .    89
Fig. 5.1      Variation in delay and loss with increasing buffer size . . . . . . .    98
Fig. 5.2      Variation in delay and loss with increasing RTS threshold. . . .    98
Fig. 5.3      Increase in delay, loss with increasing retransmissions . . . . . . .    98
Fig. 5.4      Variation of throughput with increasing retransmissions
              for various path-loss exponents. . . . . . . . . . . . . . . . . . . . . . . . .    98
Fig. 5.5      Minimal transmitter power requirement with increasing
              distance from the AP for various path-loss exponents. . . . . . . .    99
Fig. 5.6      Point of intersection for delay and loss curves obtained
              in MATLAB. . . . . . . . . . . . . . . . . . . . . . . . . . . . . . . . . . . . . . . .   100
Fig. 5.7      Delay reduction for optimal buffer size $n''$ when $delay_n$
              is within the threshold value. . . . . . . . . . . . . . . . . . . . . . . . . . .   100
Fig. 5.8      Delay reduction for optimal buffer size $n''$ when $delay_p$
              is above the threshold value . . . . . . . . . . . . . . . . . . . . . . . . . . .   101
Fig. 5.9      Loss reduction for optimal RTS threshold . . . . . . . . . . . . . . . .   102
Fig. 5.10     Flowchart of the proposed algorithm . . . . . . . . . . . . . . . . . . . .   104
Fig. 5.11     Intersection points of delay and loss in MATLAB . . . . . . . . . .   105
Fig. 5.12     Optimal buffer size and the corresponding scaled delay
              and loss plotted in MATLAB . . . . . . . . . . . . . . . . . . . . . . . . . .   106
Fig. 5.13     Received signal power with respect to distance from
              the AP . . . . . . . . . . . . . . . . . . . . . . . . . . . . . . . . . . . . . . . . . . .   106
Fig. 5.14     Optimization of delay and loss in test-bed . . . . . . . . . . . . . . .   107

Fig. 5.15    Variation of delay with increase in retransmission
             limit for different buffer sizes . . . . . . . . . . . . . . . . . . . . . . . . . .    108
Fig. 5.16    Variation of loss with increase in retransmission limit
             for buffer size of 1 Mb. . . . . . . . . . . . . . . . . . . . . . . . . . . . . . . .    109
Fig. 5.17    Variation in delay and loss for various node and AP RTS
             threshold parameters with increase in AP buffer size . . . . . . . .    109
Fig. 5.18    Selection of optimal buffer size when $n' <= n(r_{node})$
             and $delay_{node} < delay_{thresh}$. . . . . . . . . . . . . . . . . . . . . . . . .    110
Fig. 5.19    Selection of optimal buffer size when $n' <= n(r_{node})$
             and $delay_{node} > delay_{thresh}$. . . . . . . . . . . . . . . . . . . . . . . . .    111
Fig. 5.20    Selection of optimal buffer size when $n' > n(r_{node})$
             and $delay_{node} < delay_{thresh}$. . . . . . . . . . . . . . . . . . . . . . . . .    111
Fig. 5.21    Selection of optimal buffer size when $n' > n(r_{node})$
             and $delay_{node} < delay_{thresh}$. . . . . . . . . . . . . . . . . . . . . . . . .    112
Fig. 5.22    Selection of optimal RTS threshold . . . . . . . . . . . . . . . . . . . . .    112
Fig. 5.23    Optimization of delay and loss in test-bed . . . . . . . . . . . . . . .    114
Fig. 5.24    Snapshots of video calls with loss of **a** 0%, **b** 22%,
             **c** 8%, **d** 3%. . . . . . . . . . . . . . . . . . . . . . . . . . . . . . . . . . . . . .    114
Fig. 5.25    Variation of packet drop probability with increasing. . . . . . . . .    116
Fig. 5.26    Selection of threshold parameters when both delay
             and loss are within threshold . . . . . . . . . . . . . . . . . . . . . . . . . .    117
Fig. 5.27    Selection of threshold parameters when delay
             is unacceptable . . . . . . . . . . . . . . . . . . . . . . . . . . . . . . . . . . . . .    117
Fig. 5.28    Selection of threshold parameters when both delay
             and loss are above threshold. . . . . . . . . . . . . . . . . . . . . . . . . . .    118
Fig. 5.29    Variation of delay and loss for default RED threshold
             parameters in **a** uncongested medium and **b** congested
             medium. . . . . . . . . . . . . . . . . . . . . . . . . . . . . . . . . . . . . . . . . . .    119
Fig. 5.30    Variation of delay and loss for the optimized RED threshold
             parameters in **a** uncongested medium and **b** congested
             medium. . . . . . . . . . . . . . . . . . . . . . . . . . . . . . . . . . . . . . . . . . .    119
Fig. 6.1     State-space diagram for the proposed approach. . . . . . . . . . . . .    127
Fig. 6.2     Flowchart depicting the proposed approach . . . . . . . . . . . . . . .    131
Fig. 6.3     Effect of constant in/out bit rate on an ongoing call . . . . . . . . .    132
Fig. 6.4     Effect of variable bit rate on an ongoing call . . . . . . . . . . . . . .    133
Fig. 6.5     Effect of controlled load and guaranteed load service
             on packet loss in various scenarios. . . . . . . . . . . . . . . . . . . . . . .    134
Fig. 6.6     State transition diagram for the call . . . . . . . . . . . . . . . . . . . . .    136
Fig. 6.7     Variation of delay, loss, and MOS with state transitions
             in a single call scenario. . . . . . . . . . . . . . . . . . . . . . . . . . . . . . .    138
Fig. 6.8     Variation of delay, loss, and MOS with state transitions
             in a multiple call scenario. . . . . . . . . . . . . . . . . . . . . . . . . . . . .    138
Fig. 6.9     Screenshots of the video call with recorded packet loss
             of **a** 0%. **b** 3%. **c** 4%. **d** 6%. **e** 5%. . . . . . . . . . . . . . . . . . . . .    139

Fig. 6.10 Variation of delay and MOS with time.................... 143
Fig. 6.11 Variation of packet loss with time ....................... 144
Fig. 7.1 Variation in packet loss for standard codecs.............. 160
Fig. 7.2 Variation of packet loss with increase in packet payload
size........................................................ 161
Fig. 7.3 Variation of medium access delay with increase in packet
payload size ............................................ 161
Fig. 7.4 Variation in total and queuing delay with increase
in packet payload size.................................... 162
Fig. 7.5 Flowchart depicting the proposed algorithm .............. 163
Fig. 7.6 Variation in delay, loss, and MOS with time in high
bit rate scenario......................................... 164
Fig. 7.7 Variation in delay, loss, and MOS with time in low
bit rate fixed buffer scenario ........................... 165
Fig. 7.8 Variation in delay, loss, and MOS with time in low
bit rate RED buffer scenario ............................ 165
Fig. 7.9 Variation in throughput with time for low bit rate fixed
buffer scenario .......................................... 166
Fig. 7.10 Variation in throughput with time for low bit rate RED
buffer scenario .......................................... 166
Fig. 8.1 Time diagram showing jitter...................... 172
Fig. 8.2 OPNET simulation setup....................... 178
Fig. 8.3 Flowchart of the proposed adaptive jitter buffer playout
algorithm................................................ 182
Fig. 8.4 Inter-arrival jitter for network capacity of 600 kbps.
a Without playout buffer. b With proposed algorithm
(window size 100) ...................................... 184
Fig. 8.5 Inter-arrival jitter for network capacity of 1000 kbps.
a Without playout buffer. b With proposed algorithm
(window size 100) ...................................... 184
Fig. 8.6 Inter-arrival jitter for network capacity of 1400 kbps.
a Without playout buffer. b With proposed algorithm
(window size 100) ...................................... 185
Fig. 8.7 A comparison of various parameters w.r.t. window size.
a Average end-to-end delay and b packet discard ratio....... 187
Fig. 8.8 A comparison of various parameters w.r.t. window size.
a MOS. b Percentage reduction in jitter .................. 187
Fig. 8.9 Comparison of the end-to-end delays of the different
algorithms............................................... 188
Fig. 8.10 Comparison of the packet discard ratio of the different
algorithms............................................... 188
Fig. 8.11 Comparison of MOS of the different algorithms........... 189
Fig. 9.1 Packetizer/de-packetizer in a VoIP system ............... 195
Fig. 9.2 Encapsulation procedure of voice packets................ 196

Fig. 9.3      Screenshot of the voice node's attribute in OPNET .........   197
Fig. 9.4      OPNET simulation setup. ............................   198
Fig. 9.5      Increase in end-to-end delay w.r.t increase in network
              cross traffic ...................................   199
Fig. 9.6      Decrease in MOS w.r.t increase in network cross traffic ......   200
Fig. 9.7      Effect of packetization on end-to-end delay in a congested
              network. ....................................   201
Fig. 9.8      Effect of increasing number of voice sample frames
              on packetization delay. ..........................   202
Fig. 9.9      Effect of packetization on MOS in a congested network. .....   202
Fig. 9.10     Effect of packetization on 2 VoIP users with access point
              capacity of 1 Mbps. **a** End-to-end delay. **b** MOS ...........   203
Fig. 9.11     Effect of packetization on 3 VoIP users with access point
              capacity of 1 Mbps. **a** End-to-end delay **b** MOS ...........   204
Fig. 9.12     Effect of packetization on 4 VoIP users with access point
              capacity of 1 Mbps. **a** End-to-end delay **b** MOS ...........   205
Fig. 9.13     Effect of packetization on 5 VoIP users with access point
              capacity of 1 Mbps. **a** End-to-end delay. **b** MOS ...........   206
Fig. 9.14     Effect of packetization on 2 VoIP users with access point
              capacity of 2 Mbps. **a** End-to-end delay. **b** MOS ...........   207
Fig. 9.15     Effect of packetization on 3 VoIP users with access point
              capacity of 2s with access point capacity of Mbps.
              **a** End-to-end delay. **b** MOS ..........................   208
Fig. 9.16     Effect of packetization on 4 VoIP users with access point
              capacity of 2 Mbps. **a** End-to-end delay. **b** MOS ...........   209
Fig. 9.17     Effect of packetization on 5 VoIP users with access point
              capacity of 2 Mbps. **a** End-to-end delay. **b** MOS ...........   210
Fig. 9.18     Flowchart of the proposed algorithm ...................   211
Fig. 9.19     Protocol stack of the developed network node in OPNET ....   214
Fig. 9.20     Variation in packet payload size w.r.t the variation
              in end-to-end delay. ............................   214
Fig. 9.21     Improvement achieved by applying the proposed
              algorithm. ...................................   215
Fig. 9.22     Improvement in MOS achieved by applying the proposed
              algorithm. ...................................   215
Fig. 10.1     Packetizer/de-packetizer in a VoIP system ...............   218
Fig. 10.2     Concept of spectrum holes in CRN .....................   219
Fig. 10.3     WRAN architecture. ...............................   220
Fig. 10.4     Types of CR systems ..............................   220
Fig. 10.5     Relevance of the proposed research ....................   223
Fig. 10.6     Network architecture depicting deployment of VoIP
              services over CRN ..............................   229
Fig. 10.7     Flowchart depicting the proposed CR approach where SUs
              sense of channel for PU presence before transmission .......   231

Fig. 10.8    **a** Node model of VoIP over CRN for single channel scenario,
             and **b** RTP packet format in OPNET modeler 16.0.A. . . . . . . .   233
Fig. 10.9    Variation in end-to-end delay (sample mean) with sensing
             and transmission intervals of SUs. . . . . . . . . . . . . . . . . . . . . . .   234
Fig. 10.10   Variation in packet loss (maximum value) with sensing
             and transmission intervals of SUs. . . . . . . . . . . . . . . . . . . . . . .   234
Fig. 10.11   Variation in traffic received with sensing and transmission
             of SUs . . . . . . . . . . . . . . . . . . . . . . . . . . . . . . . . . . . . . . . .   235
Fig. 10.12   Variation in jitter (sample mean) with sensing
             and transmission intervals of SUs. . . . . . . . . . . . . . . . . . . . . . .   235

# List of Tables

Table 1.1   VoIP classes. . . . . . . . . . . . . . . . . . . . . . . . . . . . . . . . . . .   4
Table 1.2   Design strategy for VoIP in WiMAX. . . . . . . . . . . . . . . . . . .   17
Table 2.1   H.323 standards and protocols . . . . . . . . . . . . . . . . . . . . . . . .   26
Table 2.2   SIP Requests . . . . . . . . . . . . . . . . . . . . . . . . . . . . . . . . . . .   36
Table 2.3   Megaco commands . . . . . . . . . . . . . . . . . . . . . . . . . . . . .   41
Table 3.1   Delay specifications . . . . . . . . . . . . . . . . . . . . . . . . . . . . . .   54
Table 3.2   MOS values and their specifications . . . . . . . . . . . . . . . . . . .   58
Table 3.3   R-factor values. . . . . . . . . . . . . . . . . . . . . . . . . . . . . . . . .   59
Table 4.1   QoS metrics for SIP and H.323 . . . . . . . . . . . . . . . . . . . . . .   85
Table 5.1   Parameters in NetSim. . . . . . . . . . . . . . . . . . . . . . . . . . . . .   97
Table 5.2   Implementation details . . . . . . . . . . . . . . . . . . . . . . . . . . . .   107
Table 5.3   Voice call details with application of optimization
              techniques . . . . . . . . . . . . . . . . . . . . . . . . . . . . . . . . . . .   113
Table 5.4   RED implementation details. . . . . . . . . . . . . . . . . . . . . . . . .   120
Table 6.1   Different network scenarios . . . . . . . . . . . . . . . . . . . . . . . . .   132
Table 6.2   Delay and loss in each scenario . . . . . . . . . . . . . . . . . . . . . .   133
Table 6.3   Readings for the variable in/out bit rate in an endpoint . . . . . . .   133
Table 6.4   Different execution scenarios of RED implementation . . . . . . . .   134
Table 6.5   Heuristics for each state during the call . . . . . . . . . . . . . . . . . .   136
Table 6.6   Transition function for every link between the states . . . . . . . . .   137
Table 6.7   Category of heuristics . . . . . . . . . . . . . . . . . . . . . . . . . . . . .   137
Table 6.8   Heuristic values in multiple call scenario. . . . . . . . . . . . . . . . .   139
Table 7.1   Bandwidth versus bit rate for some popular codecs . . . . . . . . . .   152
Table 7.2   MIPS versus audio bandwidth for some popular codecs . . . . . .   153
Table 7.3   ITU-T MOS score for different codecs based on their
              compression methods [8] . . . . . . . . . . . . . . . . . . . . . . . . . . .   154
Table 7.4   Standard VoIP codecs and their parameters . . . . . . . . . . . . . . .   155
Table 7.5   Bandwidth requirements and essential codec parameters . . . . . .   158
Table 7.6   Codec parameters. . . . . . . . . . . . . . . . . . . . . . . . . . . . . . . .   160
Table 7.7   Variation of codec parameters . . . . . . . . . . . . . . . . . . . . . . . .   161
Table 8.1   Results for the algorithms for different network capacities. . . . .   179

Table 8.2   Results for the proposed algorithm for different network
            capacity with a window size of 100. . . . . . . . . . . . . . . . . . . . . .   183
Table 8.3   Effect of window size on the performance of the proposed
            algorithm for different network capacities . . . . . . . . . . . . . . . .   186
Table 9.1   Optimum number of voice sample frames to be used
            for different network scenarios. . . . . . . . . . . . . . . . . . . . . . . .   210

# Chapter 1
# Overview of VoIP Technology

## 1.1 Introduction

Technological advancements in the domain of communication have made it possible for people across the globe to stay connected among themselves. Wireless networks have further alleviated the problem of managing wired connections. The invention of mobile phones has ensured that connectivity is not hampered even while roaming from one place to another. Although the overall objective of communication (to connect every person on the earth) has not changed over the years, the focus has gradually shifted toward ensuring quality, reliability, robustness, flexibility, security, and other aspects of communication and networking.

However, the necessary infrastructure required to develop architectural models does not come without investment. As a result, communication networks were initially accessible only to the wealthy people apart from political leaders and public service departments. Gradually, the emergence of PSTN-based telephones followed by digitization of analog communication and evolution of wireless telephone technology across different generations made communication services available to the masses. Networking also witnessed a steady growth with the development of the Internet and World Wide Web, which guaranteed transmission of data from one region to another. Subsequently, Voice over IP (VoIP) was established with the integration of Internet and communication technologies in order to reduce the cost of communication and also merge the data services with voice.

Extensive research has been carried out for deploying and maintaining VoIP in practical networks, which has lead to steady rise in VoIP subscribers [1]. Widespread popularity in this domain has triggered studies on integrating VoIP with PSTN and cellular networks with the design of interfaces and gateways. Establishing VoIP as a commercial entity for the users can be attributed to two significant market strategies. The first policy is adopted by VoIP service providers as they implement VoIP as a service and develop VoIP servers, compatible phones, PBXs, gateways, etc. [2]. Some providers also create hardware- and software-based

© Springer International Publishing AG, part of Springer Nature 2019
T. Chakraborty et al., *VoIP Technology: Applications and Challenges*,
Springer Series in Wireless Technology, https://doi.org/10.1007/978-3-319-95594-0_1

modules to aid the research community in understanding the basics of VoIP tele-phony as well as for further innovations in this domain. The second strategy is followed by various application providers who integrate VoIP with their applica-tions to expand the consumer base, for example, in realistic gaming and social networking.

However, as with any technological advancement, VoIP suffers from scalability issues that are aggravated by its stringent Quality of Service (QoS) requirements [3]. This book describes those challenges faced by VoIP and discusses some novel solutions to ensure performance improvement of real-time VoIP calls even in degraded network conditions.

Voice over IP (VoIP) was established with the integration of Internet and communication technologies in order to reduce the cost of communication and also merge data services with voice.

## 1.2   What Is Voice Over IP (VoIP)?

VoIP is a technology that enables routing of voice communications through the Internet or any other Internet protocol (IP)-based networks [3]. Voice is transmitted over a general-purpose packet-switched network instead of dedicated traditional circuit-switched voice transmission lines. In order to send voice, the information has to be separated into packets just like data. Packets are chunks of information broken up into the most efficient size for routing. Thereafter, the packets need to be sent and put back together in an efficient manner. Real-time transport protocol (RTP) defines a standardized packet format for delivering audio and video over the Internet [4]. Moreover, the voice data must be compressed so that it will require less space and record only a limited frequency range. This is accomplished by suitable algorithms.

There are many protocols used in the implementation of VoIP services, the most significant being SIP and H.323 [2]. These protocols allow users to establish multimedia communication (audio, video, or other data communication) over IP networks. However, they differ significantly in design. For example, H.323 borrows heavily from legacy communication systems and is an umbrella standard com-prising of several protocols. SIP, on the other hand, does not adopt many of the information elements found in legacy systems and is an ASCII-based protocol.

VoIP is a technology that enables routing of voice communications through Internet or any other Internet protocol (IP)-based networks.

The goals of VoIP implementation are to achieve (a) significant savings in network maintenance and operations costs and (b) rapid rollout of new services. Emerging technologies such as multimedia messaging service (MMS), video calling, voicemail, and different types of VoIP services are currently being taken into use in the markets. Figure 1.1 illustrates these different means of communication.

The applications are categorized in three groups: calling, messaging, and mailing, based on the nature of the communication. Secondly, the services are differentiated on the nature of the content that is conveyed by them, i.e., text, image, voice, or video. In Fig. 1.1, the latency requirements of the services become stricter when moving from bottom to top (from mailing to calling), and the corresponding end-user value rises. On the other hand, capacity requirements increase when moving from right to left (from text to video).

A number of different types of VoIP services exist, with different players managing the required network infrastructure and servers. Differences also exist in the pricing schemes, addressing models, level of interconnection to PSTN, and mobile networks, and in the level and effects of regulation. Three fundamentally different classes of VoIP are recognized, named after an analogy to existing, well-known services and systems. The classes are PBX-like VoIP, PSTN-like VoIP, and IM-like VoIP [5, 6]. Their characteristics are summarized in Table 1.1.

PBX-like VoIP is implemented typically in large enterprises to allow low-cost communication. PSTN-like VoIP replaces the legacy PSTN-based telephone services in household and small enterprises whereas IM-like VoIP primarily targets the Internet users.

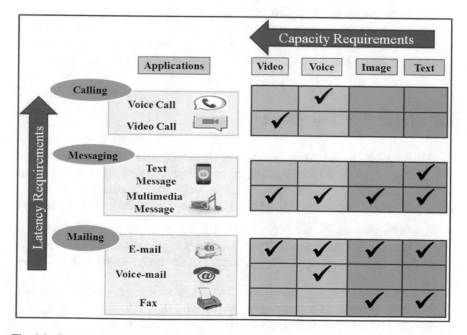

**Fig. 1.1** Communication aspects for different applications

**Table 1.1** VoIP classes

|  | PBX-like VoIP | PSTN-like VoIP | IM-like VoIP |
|---|---|---|---|
| Domain | Fixed, wireless | Fixed, wireless | Fixed, wireless, mobile |
| Target users | Large enterprises | Consumers, small businesses | Consumers |
| Managed by | Corporation (IP PBX)/local service provider (IP Centrex) | Broadband ISP/local service provider | Global service provider |
| Typical pricing scheme | Free calls inside the LAN PSTN-like pricing on outgoing calls | Free/low-cost calls to other VoIP users. PSTN-like pricing on outgoing calls | Free calls to other VoIP users. PSTN-like pricing on outgoing calls |
| QoS control | High | Medium/low | Low |
| Examples | Cisco CallManager | Vonage, Net2Phone (U.S), Ipon, Sonera | MSN Messenger, Yahoo! Messenger, Skype |

> The goals of VoIP implementation are to achieve (a) significant savings in network maintenance and operations costs and (b) rapid rollout of new services.

It must be noted in this regard that VoIP may be implemented over different networks that include wireless LAN, WiMAX, cellular networks, cognitive radio networks. Each such network has its own set of standards and regulations and introduces unique challenges that must be addressed while deploying VoIP.

## 1.3  VoIP: Why Implement It?

VoIP was initially presented as a technology that could enable a service provider to transport voice for "free" over the Internet as transport of packets over IP network was free. Gradually, it found applications in residential telephony as well as in office networks. The success of VoIP can be attributed to the following key reasons.

- *Ease of deployment*—Many functions requiring multiple distributed points of presence can be centralized in VoIP domain owing to the unique characteristics of VoIP call-controllers, thereby reducing administrative overheads and accelerating deployment.
- *Simplification of transport networks*—Standard IP networks after proper configuration can be used to carry VoIP packets, thus eliminating the need to establish leased lines dedicated to voice prior to establishment.

- *Cost reduction*—There is significant a reduction in operational and maintenance costs. This is especially beneficial for companies that actively make a lot of calls on a daily basis, or for those who execute long-distance international calls.
- *Value-added services*—VoIP infrastructure can be utilized to host and implement various services for consumers like MMS, PTT.
- *Anytime, anywhere communication*—IM-based VoIP offers anytime, anywhere communication to customers having access to Internet and a registered account, thereby removing the problems of infrastructure-based modes of communication.
- *Easy upgradation*—VoIP services can be easily upgraded owing to the simplicity of VoIP operations.

However, as VoIP operates over IP which is the "best-effort" protocol, it requires certain QoS guarantees [3]. Real-time loss-sensitive nature of voice communication implies that extreme care must be taken toward maintaining the quality of call that is acceptable to the end user. Security concerns should also be addressed since voice packets can easily be compromised and private sessions hacked.

> VoIP could enable a service provider to transport voice for "free" over the Internet as transport of packets over IP network is free. However, as VoIP operates over IP which is the "best-effort" protocol, it requires certain QoS guarantees.

## 1.4   Evolution of VoIP

There are two fundamental technologies that are necessary for the existence of VoIP, namely telephone and Internet. Telephony has its origin with telegraphy in 1844, when Samuel Morse developed the capability to send pulses of electric current over wires that spanned distances farther than one could shout, walk, or ride. Voice communication became possible with the invention of telephone by Graham Bell on March 10, 1876. By 1906, American inventor, Lee De Forest, invented a three-element vacuum tube that revolutionized the entire field of electronics by allowing amplification of signals, both telegraphy and voice. Wireless voice communication using amplitude modulation (AM) was realized during the 1920s. The ensuing years saw tremendous growth in radio station broadcasting that brought the possibility of real-time information to the public. Of course, wires still had their place because radio was not always the most reliable medium due to environmental factors.

Telephone technology progressed steadily, and telegraphy still found a place in data communications in the form of the telegram. Radio technology advanced throughout in the 1930s with the notable invention of frequency modulation (FM), which provided better sound quality and was more resistant to interference than the older AM broadcasting system. The post-World War II period saw an explosion of innovations with the development of the transistor (December 1947) and the birth of the computer. Computers provided a tool for people to process and transfer lots of data at high speed. The Space Age began with the launch of the Soviet satellite Sputnik on October 4, 1957. Satellite communications provided reliable long-distance communications by augmenting or replacing cables. This created the demand for reliable, anytime, anywhere communications. However, it took almost 26 years after Sputnik before cellular communications brought mobile voice communications to the masses.

In the early days of telephony, whenever a user wanted to talk to another person, they would ring the operator and give the name or number of the other party. Next, the operator would connect a patch cord (2-wire cable with a jack plug on each end) between the two phones and the two people could communicate. Bundles of wires called trunks ran between exchanges, forming proto-networks. Networks were connected together hierarchically until they connected countries across the world. This was the beginning of the Public Switched Telephone Network (PSTN) which is now the worldwide collection of interconnected public telephone networks designed primarily for voice traffic. It is a circuit-switched network where a dedicated circuit is established for the duration of a transmission, such as a telephone call. Originally only an analog system, the PSTN is now almost entirely digital and employs efficient algorithms to ensure the reliability of voice calls.

Parallel growth in networking ensured the development of Internet in 1968, by ARPANET, which was followed by the design of hypertext transfer protocol (HTTP) and hypertext markup language (HTML). This marked the advent of the World Wide Web (WWW), which increased the popularity of Internet. The transmission control protocol/Internet protocol (TCP/IP) was created in the year 1971 by Dr. Vint Cerf that defined the nature of data packets to be sent via Internet and established the rules for routing of packets to their destinations. With the rapid growth of Internet and deregulation of the telecommunications industry, infrastructure convergence in the form of building voice applications on top of data networks gained momentum. This is suitably illustrated in Fig. 1.2. Eventually, this led to the birth of VoIP which allowed voice to be carried by IP packets over any IP-based networks such as Internet.

With the rapid growth of Internet and deregulation of the telecommunications industry, infrastructure convergence in the form of building voice applications on top of data networks led to the birth of VoIP.

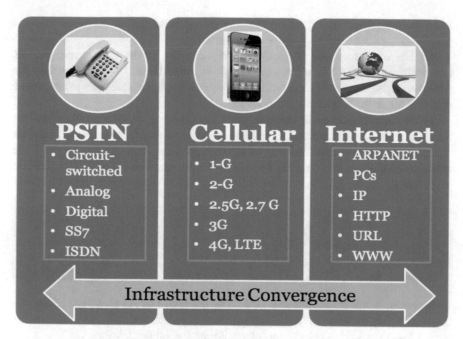

**Fig. 1.2** Evolution of VoIP technology

## 1.5   How VoIP Works?

VoIP communication is characterized by telephone networks that can be inter-connected with Internet. The basic technology consists of digitizing the analog voice and sending it in the form of IP packets over the Internet or any other IP-based network. An audio input device, such as a microphone, is required at the sending end. The audio signal as recorded by the input device is sampled at a very high rate (at least 8000 times per second or more) and transformed into digital form by an analog-to-digital (A/D) converter. The digitized data is further compressed into very small samples that are collected together into larger chunks and placed into data packets for transmission over the IP network. This process is referred to as packetization. Generally, a single IP packet will contain 10 or more milliseconds of audio, with 20 or 30 ms being most common. There are a number of ways to compress this audio, the algorithm for which is referred to as a "compressor/de-compressor", or simply Codec [7]. Many Codecs exist for a variety of applications (e.g., movies and sound recordings). With respect to VoIP, the Codecs are optimized for compressing voice, which significantly reduces the bandwidth used compared to an uncompressed audio stream and ensures high quality of VoIP transmission. Most of the Codecs are defined by standards of the International Telecommunication Union, the Telecommunication division (ITU-T). Each of them has different properties regarding the amount of bandwidth it requires and the perceived quality of the encoded speech signal.

After binary information is encoded and packetized at the sender end, packets encapsulating voice data can be transmitted on the network. Figure 1.3 shows the end-to-end path as needed for VoIP communication (a similar path exists in the opposite sense for a bidirectional connection).

Voice packets interact in the network with other application packets and are routed through shared connections to their destination. At the receiver end, they are decapsulated and decoded. The flow of digital data is then converted to analog form again and played at an output device, usually a speaker. The entire communication is illustrated in Fig. 1.4.

> The basic technology of VoIP consists of digitizing the analog voice and sending it in the form of IP packets over the Internet or any other IP-based network.

However, some IP packets can be lost in the network. As real-time communication is highly sensitive to loss of information, steps must be taken to minimize the packet loss through the reservation of resources and other techniques. Codecs can compensate for these lost packets by "filling in the gaps" with audio that is acceptable to human ear. This process is referred to as packet loss concealment

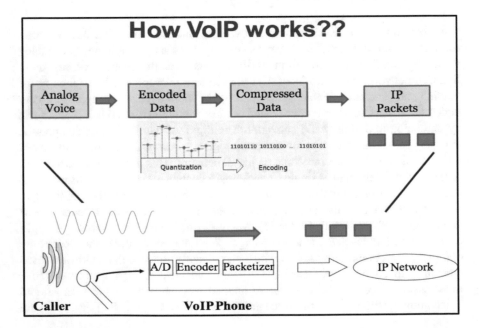

**Fig. 1.3** Working mechanism behind VoIP communication

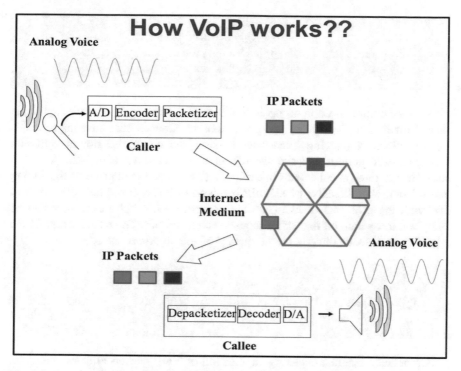

**Fig. 1.4** Complete VoIP communication between the sender and the receiver

(PLC) [3]. Redundancy is another strategy where packets are sent multiple times in order to overcome packet loss. Error recovery techniques like forward error correction (FEC) include some information from previously transmitted packets in subsequent packets. Thereafter, the lost packet is reconstructed from the information bits in neighboring packets by applying suitable mathematical operations in a particular FEC scheme. Packets considered to be lost may actually reach the destination after incurring a significant amount of delay. It is very common for applications to experience out-of-order packets in a packet-switched network. This is particularly problematic for VoIP systems, as delays in delivering a voice packet means the information is too old to play. Such old packets are discarded in small amounts by the PLC algorithms and perceived as being lost in the network.

Packet loss and delay can be calculated from the information in the IP packets. During the course of conversation, such delays can increase and decrease, resulting in a variable delay component referred to as jitter. While delay has to be maintained under the threshold level, jitter can result in choppy voice or temporary glitches and must also be minimized. Jitter buffer algorithms are implemented by VoIP applications in this regard. A certain number of packets are queued before play-out, and the queue length may be increased or decreased over time to reduce the number of discarded, late-arriving packets or to reduce the "mouth to ear" delay.

As real-time communication is highly sensitive to loss of information, steps
must be taken to minimize the end-to-end delay and packet loss through
reservation of resources and other techniques.

As voice communication occurs in the form of talkspurts [8], there is significant
"dead" time during which no speaker is talking. Codecs take advantage of this
silence periods by applying "silence suppression" techniques that stop transmission
during the idle periods and significantly save the network bandwidth. "Comfort
noise" is generated at either ends to ensure that the users do not perceive the drop in
transmission. Additionally, when VoIP is linked to PSTN-based networks, it must
deal with the problems of line echo and acoustic echo. Echo cancellers can be
optimized to operate on line echo, acoustic echo, or both. The effectiveness of the
cancellation depends directly on the quality of the algorithm used.

Codecs take advantage of the silence periods by applying "silence suppression" techniques that stop transmission during the idle periods and significantly save the network bandwidth.

A schematic diagram depicting the creation of VoIP packets after compression,
echo cancellation, and silent suppression is provided in Fig. 1.5. The input consists
of time-slotted PCM bit stream that is converted into VoIP packets at the output end.

The successful deployment of VoIP application must also address several other
concerns apart from just sending and retrieving the audio/video packets over the
Internet. There must be an agreed call signaling protocol to initiate, manage, and
terminate the call sessions. Protocols must be in place to allow the computers to find
each other and decide on the nature of information exchange before packet flow is
initiated. There should also be an agreed format (packet payload format) for the

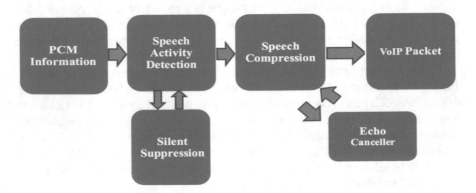

**Fig. 1.5**  Development of VoIP packet

contents of the media packets. In addition to computers, VoIP can also be implemented in mobile IP-based phones, analog terminal adapters (ATAs), and gateways. Each such design entails different aspects of communication and networking that must be investigated.

> The successful deployment of VoIP application must address several other concerns apart from just sending and retrieving the audio/video packets over the Internet, such as the call signaling protocols. e.g., SIP, H.323.

An architectural setup for VoIP-based communication system comprising of the different entities is demonstrated in Fig. 1.6 that depicts the deployment of VoIP technology in various IP and PSTN-based networks.

The fundamental elements required to deploy VoIP in a public network are enlisted as follows.

1. IP-enabled workstation—The end users that are involved in VoIP communication must have IP-enabled devices that are compatible with VoIP protocols and allow routing through IP-based networks. The workstation can be a soft phone installed in a computer with access to an IP network. IP-enabled mobile and fixed telephones can also be used to implement VoIP technology.

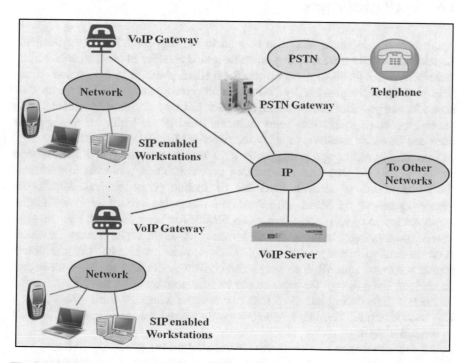

**Fig. 1.6** Fundamental elements of a VoIP Network

2. VoIP server—The VoIP server is the centralized node that initiates, manages, and terminates communication between the caller and the callee. (in telephony terms, caller is the user who initiates the call and a callee receives the call at the other end). The VoIP server must implement the call signaling protocols (SIP, H.323, etc.) and ensure proper routing of IP packets to their destination. Call admission control is one of the primary functions of the server. It can also be used for QoS provisioning mechanisms.

3. Gateway—One way to increase the interoperability of VoIP is by implementing it in diverse networks having different characteristics which is made possible using gateways. Gateways ensure proper coordination in between these networks and further allow VoIP users to communicate to PSTN-based telephones. Moreover, firewalls can be implemented in gateways to achieve secured communication with appropriate packet filtering rules.

4. Gatekeeper—A gatekeeper is a management tool that oversees authentication, authorization, telephone directory, and PBX services. Commercial entities implementing VoIP can maintain the billing information along with the call details in the gatekeeper. Although it can be incorporated in the server, generally gatekeepers are implemented separately to simplify the server operations.

## 1.6  VoIP Applications

Apart from IP telephony, VoIP can be used to develop several other applications. As telecom operators are focusing on value-added services in order to attract more customers, VoIP technology serves as an excellent platform to design such tools. This is only made possible by "infrastructure convergence" of VoIP with data networks and provides a low-cost easy solution to many real-world problems. With the emergence of simple text-based protocols like SIP for VoIP, existing applications can be easily modified to enjoy the benefits of all-IP communication.

One of the most popular applications in mobile telephony is short message service (SMS) [9] that uses standardized communication protocols to exchange short text messages between fixed line or mobile phone devices. Multimedia messaging service (or MMS) [9] extends the messaging service to allow simultaneous exchange of text, audio, and video files. It has been observed that there are several disadvantages of traditional GSM-based messaging. The primary drawback is on the limit of messages that a single GSM modem can handle. The situation is further worsened with MMS messages that contain media files and consume higher bandwidth. Even though the capacity can be increased by using more modems, the connection fails during increased traffic activity as witnessed during New Year's Day and other public holidays. Moreover, only a fixed sender address can be used to send messages.

All these problems can be effectively reduced by implementing IP-based messaging using the existing VoIP infrastructure. Firstly, any IP-enabled device can send the messages. Thus, users can transmit messages during emergency even in the absence of a phone or GSM connection. Moreover, low signal strength would not affect IP message delivery. Finally, the same IP device can be used by different senders to forward messages using their own registered accounts. As per industry reports, the mobile industry reported losses of over US$ 10 billion due to declining text messages sent by users as they switched to IP-based messaging service.

Multiconferencing applications are also relying on VoIP so that professionals from across the business and service sector can instantly connect among themselves "on the fly." In addition to the cost-saving benefits derived from using VoIP conferencing, users also can take advantage of its additional facilities such as checking the voicemail over the Internet, attaching messages to email, and sharing files during conversation. An overview of the VoIP-based multiconferencing system is depicted in Fig. 1.7. The VoIP conference server performs authentication and authorization to allow only registered users in the conference. Call details along with the necessary statistics are also gathered for accounting purposes.

There has also been a tremendous growth in social networking applications with emergence of Facebook, LinkedIn, Gtalk, and other popular Web-based services that allow people from and around the globe to connect and share ideas. These Web sites provide users with their registered accounts to meet other people having similar interests and interact via text, audio, and video chats. As most of these service providers operate without charging any money from the users, VoIP is the best option to implement the required facilities.

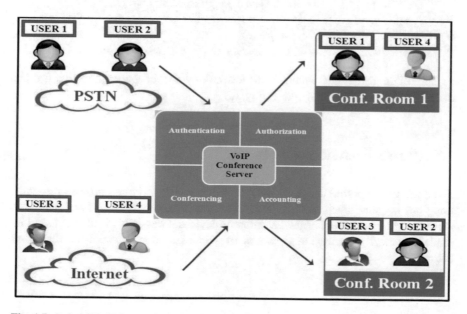

**Fig. 1.7** Role of VoIP in multiconferencing system

Multiplayer gaming is another domain where VoIP has a bright prospect [10]. Games, nowadays, give players realistic models, believable backgrounds, and the ability to use the Internet to connect to millions of other gamers all over the world. Allowing multiplayer gamers to cooperate and play has been made possible with VoIP, which, despite its shortcomings, has made such games more exciting. The primary advantage is the seamless integration of the gaming software with VoIP applications, thereby allowing rapid immersion where users do not have to switch to separate windows to share information with fellow gamers. VoIP also enables users to invite players into the game at runtime without any interruption, thereby retaining the excitement level in these games and expanding the popularity of the gaming applications.

Currently, military organizations are also transitioning their telephony infrastructure from legacy TDM to next-generation networks (NGN) based on VoIP technology [11]. Other than the basic advantages of all-IP-based VoIP communication, VoIP proves to be more resilient than TDM networks and easier to manage compared to their older TDM counterparts. Security and survivability are obvious military requirements that must be satisfied by VoIP applications. Accordingly, military VoIP networks are based on modified versions of the standard protocols. One example is MLPP that is an ITU-defined standard being used by US Department of Defense and provides a prioritized call handling service having multilevel precedence and preemption features.

> VoIP continues to mark its presence in almost every sector that involves IP-enabled services and is going to enjoy a significant share of the total Web-based traffic.

The different domains where VoIP technology can be put to effective use are endless and can be summarized in Fig. 1.8.

## 1.7  VoIP: Present and Future

Since the first reported usage of VoIP in 1995, the underlying software programming has become much more refined. The protocols have been updated for better networking and compatibility with different devices. Gradually, VoIP has found its implementation in emerging networks of wireless communication as discussed below.

# VoIP Applications

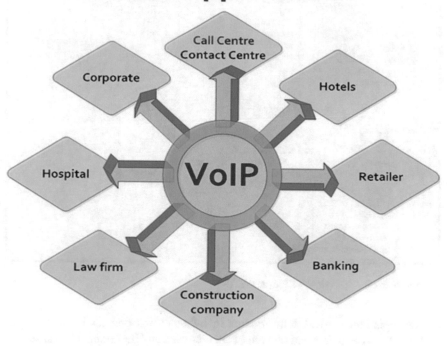

**Fig. 1.8**  Application areas for VoIP

## 1.7.1  VoIP Over WLAN

Wireless LANs have rapidly become pervasive among enterprises. It is considered as the backbone for office and campus networks. Introduction of new standards in WLAN (IEEE 802.11 a/b/g/n) [12] and availability of wireless voice clients have made WLAN one of the most widely deployed networks. Arrival of dual-mode smartphones enabling both wireless and cellular connectivity has expanded the subscriber base of WLANs. Accordingly, VoIP over WLAN (commonly referred to as VoWLAN) has a huge prospect.

There are several design challenges for VoWLAN. The real-time constraints must be addressed in accordance with the standards implemented in WLAN. As WLAN is based on a random access protocol that allows clients to roam freely and is operated in the unlicensed ISM band, there are numerous design implications for VoIP implementation that require careful analysis and planning, such as the following.

- maintaining Quality of Service for wireless "over the air" links through appropriate reservation of resources and other QoS policies

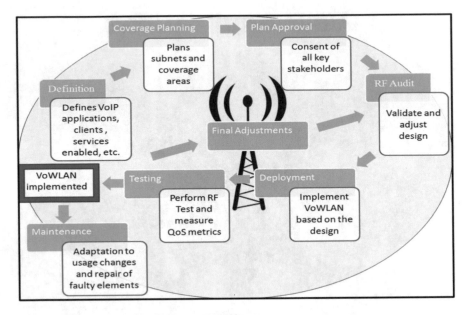

**Fig. 1.9** Steps for deploying VoIP over WLAN

- developing the network infrastructure such that VoIP server can interact with the WLAN access points and other network elements and designing of local subnets to ensure maximum coverage
- formulating policies to implement network security including authentication of the users and encryption of the information bits
- integrating software modules for VoIP with WLAN and ensure its compatibility with the operating system of the VoIP clients.

The entire process of incorporating VoIP in WLAN is carried out in several steps as illustrated in Fig. 1.9.

> As WLAN is based on a random access protocol that allows clients to roam freely and is operated in the unlicensed ISM band, there are several design implications for VoIP implementation that require careful analysis and planning.

### 1.7.2  VoIP Over WiMAX

WiMAX refers to interoperable implementations of the IEEE 802.16 family of wireless network standards ratified by the WiMAX Forum [13]. A WiMAX-based network achieves a data rate of almost 30–40 megabits per second for fixed and portable access applications. Mobile WiMAX further enables the convergence of mobile and fixed broadband networks through a common wide area broadband radio access technology and flexible network architecture. With higher throughput, coverage, and data rate as compared to Wi-Fi, WiMAX serves as an effective platform to host VoIP technology and perform multimedia communications [14].

The WiMAX 802.16e (mobile WiMAX) standard provides five Quality of Service (QoS) classes in the air interface (between the access point and mobile subscriber) that allow for different traffic flows on the network. Unsolicited grant service (UGS) supports constant bit rate (CBR) services and can be used for VoIP applications without silent suppression. Real-time polling service (rtPS) enables streaming of audio and video. Extended rtPS combines the characteristics of UGS and rtPS and supports VoIP with silent suppression.

In order to design a voice-capable WiMAX network and enable customer-centric VoIP service, a service provider must devise a competitive strategy keeping in mind the customer requirements and the pricing strategy [15]. The design plan for VoIP in WiMAX is demonstrated in Table 1.2.

However, there are several design challenges before VoIP can effectively be utilized by WiMAX users. Till date, only a few laptops and computers are equipped with WiMAX reception cards, and even lesser number of mobile phones support WiMAX. WiMAX modems exist on the market, which can be used to interface with the computers and phones. However, due to their power requirements, they adversely affect mobility. An efficient WiMAX connection should be accompanied with good network management having adequate facilities, both on the hardware and software side, so as to incorporate VoIP applications and maintain the QoS of VoIP communication. Moreover, as VoIP is characterized by small-size, periodic packets, supporting a large number of simultaneous VoIP users in a WiMAX network is a challenge due to "persistent scheduling" approach adopted by WiMAX for VoIP services.

With higher throughput, coverage, and data rate as compared to Wi-Fi, WiMAX serves as an effective platform to host VoIP technology and perform multimedia communications. However, there are several design challenges before VoIP can effectively be utilized by WiMAX users.

**Table 1.2**  Design strategy for VoIP in WiMAX

|  | Residential | Small/medium business | Mobile customer |
|---|---|---|---|
| Service as per user requirements | Maintain moderate voice quality and enable basic features | Maintain high voice quality and enable basic features | Maintain basic voice quality and enable high features |
| Pricing | Low | High | Competitive |
| Strategy planning | Cost-effective solution for voice and data | Offer better product for maximum user satisfaction | Overcome the difficulties of mobility and offer basic service |

### 1.7.3   Voice Over LTE

Cellular telephony has recorded a rapid increase in users and data traffic with the inclusion of several value-added services, which demands for increased data transmission speeds and lower latency. This has resulted in continuous development of cellular technology as it witnessed transition from 1G to 3G networks. 3G networks have further lead to the emergence of all-IP-based LTE (Long-Term Evolution) networks [16]. In contrast to 3G networks' usage of circuit-switched voice and SMS and packet-switched data, LTE deploys a complete IP-based network infrastructure with the help of Evolved UMTS Terrestrial Radio Access Network (eUTRAN) and Evolved Packet Core (EPC), thereby providing improved bandwidth and higher QoS.

Several solutions have been proposed for transfer of voice over LTE that includes circuit-switched fall back (CSFB), simultaneous Voice over LTE (SV-LTE), Voice over LTE via GAN (VoLGA). However, the most efficient approach is the IMS (IP Multimedia System)-based Voice over LTE (VoLTE) that provides VoIP and SMS in LTE using a fully packet-switched network, and is a 3GPP standard for LTE voice [17]. IMS VoIP is based on a 3GPP standardized implementation of SIP and solves the challenges of retaining LTE data rates. It also minimizes the excessive voice call setup delay and provides a flat, all-IP network for operational savings and HD voice, and creates the platform for new IP-based services.

> The most efficient approach is the IMS (IP Multimedia System)-based Voice over LTE (VoLTE) that provides VoIP and SMS in LTE using a fully packet-switched network, and is a 3GPP standard for LTE voice.

### 1.7.4   VoIP Over Cognitive Radio Network

Cognitive radio network is one of the most promising technologies in the domain of wireless networks, which focuses on increasing bandwidth utilization through opportunistic mode of communication [18]. A set of licensed (primary) users occupy the available spectrum band. Once they are idle, unlicensed secondary users are allowed to access those spectrum bands. However, care must be taken to minimize the interference between the primary and secondary users, which involves several steps including spectrum sensing, spectrum analysis, management, mobility.

The primary objective of cognitive radio network is to reduce spectrum congestion and utilize the bandwidth at the maximum. At the same time, considering the immense popularity of VoIP and its phenomenal growth in the coming years, supporting more number of VoIP users in a network with limited radio resources while maintaining adequate QoS for every VoIP call becomes impossible. Therefore, deployment of VoIP applications in such next-generation CRN is a promising aspect and is being thoroughly investigated in research communities [19].

However, VoIP communication by unlicensed users in such unpredictable networks result in severe degradation of VoIP call quality unless adequate QoS strategies are defined to guide VoIP traffic in such active networks. As cognitive radio network spans several other networks, ensuring compatibility of the protocols and related software modules in VoIP is another major problem that can only be solved by necessary architectural modifications and design of suitable interfaces. An overview of the network infrastructure for VoIP in cognitive radio network is provided in Fig. 1.10.

> Considering the immense popularity of VoIP and its phenomenal growth in the coming years, supporting more number of VoIP users in a network with limited radio resources while maintaining adequate QoS for every VoIP call becomes impossible. That is where cognitive radio networks can prove highly effective.

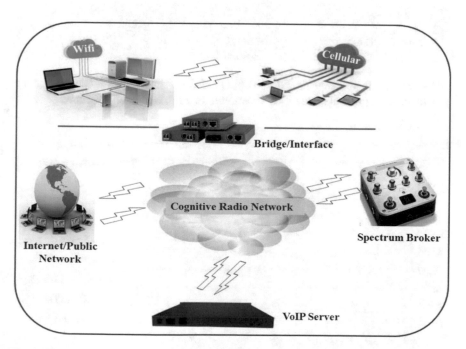

**Fig. 1.10** Architectural overview of VoIP in cognitive radio network

## 1.8   VoIP Popularity

The advantages of VoIP can only be realized through increased usage of Internet. As per ITU-T, recent statistics record increased usage of Internet by consumers across the globe that reaches almost 90%, thereby implying the huge prospect of Internet-based VoIP services in these countries. Even developing countries like India and Malaysia have witnessed greater utilization of Internet owing to higher literacy rates and increased availability of Internet-based connectivity [20] as depicted in Fig. 1.11.

Additionally, the total number of mobile subscribers has increased drastically in all countries, more specifically in the developing sector. At the same time, the number of people opting for mobile broadband subscriptions has increased manifold over the past few years [20] as depicted in Fig. 1.12. This trend highlights the growing dominance of mobile-based IP services and underlines the significance of the VoIP as the primary communication medium in this domain.

Quite obviously, the popularity of VoIP both as a service and an application has gained immense momentum in the recent years, resulting in high volumes of VoIP traffic generated [20] as shown in Fig. 1.13. The industry success is also stimulated by the US Federal Communication Commission decision not to control or limit voice traffic over the Internet, and the cheap price of these services. With low barriers to entry, competition is growing making companies differentiate services, improve quality, and reduce prices.

> With low barriers to entry, competition is growing making companies differentiate services, improve quality and reduce prices.

As per reports published in Point Topic [21], VoIP has witnessed a phenomenal growth in terms of subscribers as illustrated in Fig. 1.14 with the emergence of

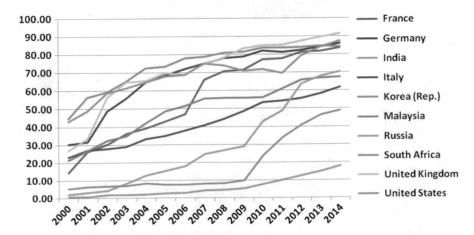

**Fig. 1.11** Percentage utilization of Internet by users in different countries

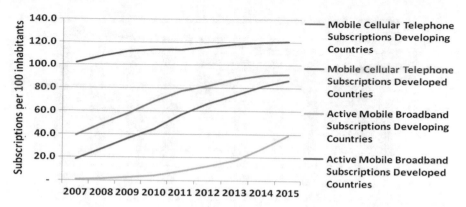

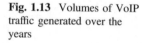

**Fig. 1.12**  Mobile telephony and mobile broadband users in developing and developed countries

**Fig. 1.13**  Volumes of VoIP traffic generated over the years

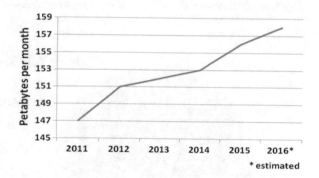

professional VoIP services along with increased deployment in campus and office networks. The primary reason behind this surge is not only the cost-effective solution of VoIP but also includes choosing the right telephony solution that will provide "exceptional customer experience" and boost its perception, which is where VoIP excels over traditional telephony solutions. Moreover, increasing number of households especially in the developed countries of USA and Europe are switching to VoIP service from traditional landline telephony due to its easy integration with other interactive services.

The continent-wise distribution of the VoIP subscribers is depicted in Fig. 1.15. It is evident that VoIP services are most efficiently utilized in the highly developed areas of North America and Europe. Countries like Korea Republic and Japan in the East Asia also enjoy a significant share of the overall VoIP usage. Specifically, the USA became the country with the highest number of subscribers in 2013. During the same period, France recorded the maximum percentage growth as almost 95% of the overall number of fixed broadband users in the country subscribed for VoIP.

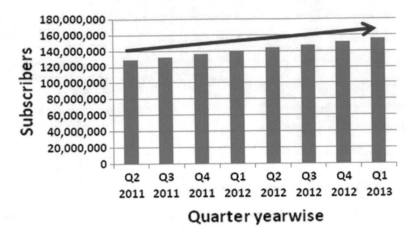

**Fig. 1.14** Total number of VoIP subscribers for various quarters of 2011–2013

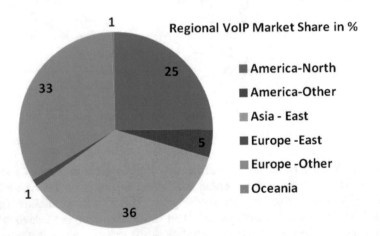

**Fig. 1.15** Distribution of VoIP users across continents

The primary reason behind this surge is not only the cost-effective solution of VoIP but also includes choosing the right telephony solution that will provide "exceptional customer experience" and boost its perception, which is where VoIP excels over traditional telephony solutions.

## 1.9 Summary

Voice over IP (VoIP) is a good solution used to reduce the cost of communication and merge data services with voice. Courtesy to the development of telephone and Internet, VoIP could be implemented to achieve (a) significant savings in network maintenance and operations costs and (b) rapid rollout of new services. The technology involves digitizing the analog voice and sending it in the form of IP packets over the Internet.

Benefits of VoIP include the ease of deployment, use of standard IP networks already in existence to carry its packets, significant reduction in cost, easy upgradation, and more.

Some industries that can use VoIP technology are corporates, hospitals, hotels, banking sector, law firms, construction companies, etc.

## References

1. White Paper on The Future of VoIP Interconnection (GSR, ITU-T, 2009)
2. C. Vaishnav, Voice over Internet Protocol (VoIP): The Dynamics of Technology and Regulation, M.S. Thesis, Massachusetts Institute of Technology (2006)
3. International Telecommunication Union (ITU), The Status of Voice over Internet Protocol (VOIP) Worldwide, 2006, New Initiatives Programme, Document: FoV/04, 12 Jan 2007
4. H. Schulzrinne, S. Casner, R. Frederick, V. Jacobson, RTP: A Transport Protocol for Real-Time Applications, RFC-1889, Jan 1996
5. T. Smura, H. Hämmäinen, The role of VoIP: future evolution paths of voice communication, in *Proceedings of 1st International CICT Conference*, Denmark, 2004
6. J. Davidson, B. Gracely, J. Peters, Overview of the PSTN and Comparisons to Voice over IP, Book Chapter (CISCO Press, 1 Jan 2001)
7. B. Goode, Voice over Internet Protocol (VoIP), in *Proceedings of the IEEE*, vol. 90, no. 9, pp. 1495–1517, Sept 2002. https://doi.org/10.1109/jproc.2002.802005
8. J. Wenyu, H. Schulzrinne, Analysis of on-off patterns in VoIP and their effect on voice traffic aggregation, in *Proceedings of Ninth International Conference on Computer Communications and Networks*, pp. 82–87, 16–18 Oct 2000, USA, https://doi.org/10.1109/icccn.2000.885474
9. 3rd Generation Partnership Project; Technical Specification Group Services and System Aspects; Support of SMS and MMS over generic 3GPP IP access (Release 7), Technical Report, 3GPP TR 23.804 V7.1.0 (2005–09), 2005
10. G. Nugent, Voice over Internet Protocol (VoIP), Video Games, and the Adolescent's Perceived Experience, Ph.D. Thesis, Walden University, Jan 2015
11. G. Fromentoux, R. Moignard, C. Pageot-Millet, A. Tarridec, IP networks towards NGN, in *Proceedings of IP Networking & Mediacom-2004* ITU-T Workshop, Apr 2001
12. White Paper on Design Principles for Voice over WLAN (CISCO Press, 2007)
13. WiMAX Forum Network Architecture, WiMAX Forum Network Architecture—Stage 3—Detailed Protocols and Procedures—WiMAX Forum Document Number WMF—T33-001-R010v04, WiMAX Forum, 03 Feb 2009
14. M. Atif Qureshi et al., Comparative study of VoIP over WiMax and WiFi. IJCSI Int. J. Comput. Sci. Issues **8**(3), 433–437 (2011)
15. Solution Paper on WiMAX VoIP: How Motorola Can Help You Offer a Successful Service (Motorola, 2007)

16. Strategic White Paper on The LTE Network Architecture (Alcatel-Lucent, 2009)
17. Strategic White Paper on Voice over LTE: The New Mobile Voice (Alcatel-Lucent, 2012)
18. F. Akyildiz, W.Y. Lee, M.C. Vuran, S. Mohanty, Next-generation/dynamic spectrum access/ cognitive radio wireless networks: a survey. Comput. Netw. J. **50**, 2127–2159 (2006). https:// doi.org/10.1016/j.comnet.2006.05.001. (Elsevier)
19. T. Chakraborty, I.S. Misra, T. Manna, Design and implementation of VoIP based 2-tier cognitive radio network for improved spectrum utilization. IEEE Syst. J. (Early Access), 23 Jan 2015. https://doi.org/10.1109/jsyst.2014.2382607
20. International Telecommunication Union (ITU), ICT Facts and Figures, Geneva, May 2015
21. Point Topic, VoIP Statistics—Market Analysis, June 2013

# Chapter 2
# VoIP Protocol Fundamentals

## 2.1 Introduction

Successful VoIP communication is ensured by implementing robust protocols that not only manage to preserve the QoS of ongoing transmissions but also ensure maximum utilization of the system resources. Signaling protocols play a significant role in this aspect and are used to establish and control multimedia sessions. These sessions include multimedia conferences, telephony, distance learning, and similar applications. The IP signaling protocols are used to connect software- and hardware-based clients through a local area network (LAN) or the Internet.

The primary functions of call signaling protocols in the domain of VoIP communication are establishing calls and controlling sessions using features like user location lookup, name and address translation, connection setup, feature negotiation, feature change, call termination, and call participant management such as invitation of more participants. Additional services like security, billing, session announcement, and directory services can also be included in the protocols. It must be clarified that although signaling is closely related to the transmitted data streams, data transmission is not a part of the signaling protocols.

There are currently two standardized protocols widely deployed in the market, namely H.323 and SIP. These two protocols provide different approaches toward attaining the same goal—signaling and control of multimedia conferences. H.323 was approved in 1996 and is an umbrella standard from the International Telecommunications Union (ITU) for multimedia communications over long-distance networks [1]. The more recent Session Initiation Protocol (SIP) is developed by the Multiparty Multimedia Session Control (MMUSIC) working group of the Internet Engineering Task Force (IETF) [2]. Unlike H.323, SIP is a much more lightweight protocol based on HTML.

© Springer International Publishing AG, part of Springer Nature 2019
T. Chakraborty et al., *VoIP Technology: Applications and Challenges*,
Springer Series in Wireless Technology, https://doi.org/10.1007/978-3-319-95594-0_2

Signaling protocols are used to establish and control multimedia sessions. There are currently two standardized protocols widely deployed in the market, namely H.323 and SIP. These two protocols provide different approaches toward attaining the same goal—signaling and control of multimedia conferences.

## 2.2   H.323

H.323 is an ITU-T specification for transmitting audio, video, and data across an IP network, including the Internet. It is an umbrella specification that covers many other ITU documents and protocols. It provides a complete specification of the architecture required to deploy voice and videoconferencing systems over a packet network. The H.323 standard addresses call signaling and control, multimedia transport and control, and bandwidth control for point-to-point and multipoint conferences. Accordingly, major VoIP service providers have increasingly used its services to establish and manage VoIP sessions. Since H.323 version 4, it has been possible for any organization to develop extensions to the H.323 protocol to enable new functionality [1].

The H.323 standard consists of the components and protocols as mentioned in Table 2.1.

H.323 uses the Abstract Syntax Notation (ASN.1) syntax and that is the reason for its reputation of being a "complex" protocol. ASN.1 is defined in ITU X.680 and its serialization [actual bit-level representation of the structured data for transport over a network used to code H.323 protocol data units (PDUs)] is defined in ITU X.691. ASN.1 defines two ways of serializing data for transport over a network, namely Basic Encoding Rules (BER) and Packet Encoding Rules (PER).

H.323 is an umbrella specification that covers many other ITU documents and protocols and is used for transmitting audio, video, and data across an IP network, including the Internet.

**Table 2.1** H.323 standards and protocols

| Feature | Protocol |
|---|---|
| Call signaling | H.225 |
| Media control | H.245 |
| Audio codecs | G.711, G.722, G.723, G.728, G.729, etc. |
| Video codecs | H.261, H.263, etc. |
| Data sharing | T.120 |
| Media transport | RTP/RTCP |

## *2.2.1   Elements*

H.323 elements play a pivotal role in maintaining efficient VoIP service. The various elements are discussed as under.

### Terminal

H.323 terminals must have a system control unit, media transmission, audio codec, and packet-based network interface. Optional requirements include video codec and user data applications. It provides H.225 and H.245 call control, capability exchange, messaging, and signaling of commands for proper operation of the terminal. It also formats the transmitted audio, video, data, control streams, and messages onto network interface and receives these messages from the network interface.

### Gateway

The H.323 gateway reflects the characteristics of a switched circuit network (SCN) endpoint and H.323 endpoint. It translates into audio, video, and data transmission formats as well as communication systems and protocols. This includes call setup and teardown on both the IP network and SCN. It also performs compression and packetization of voice.

### Gatekeeper

An optional entity, the gatekeeper, provides pre-call and call-level control services to H.323 endpoints. Gatekeepers are logically separated from the other network elements in H.323 environments. It performs address translation, admissions control, bandwidth control, and zone management.

### Multipoint Control Unit (MCU)

MCUs consist of multipoint controller and multipoint processor. They are involved in issues related to conferencing.

Figure 2.1 shows the various H.323 elements and their interactions.

## *2.2.2   Protocol Suite*

The H.323 protocol suite is based on several protocols, as illustrated in Fig. 2.2. The protocol family supports call admissions, setup, status, teardown, media streams, and messages in H.323 systems. These protocols are supported by both reliable and unreliable packet delivery mechanisms over data networks [3].

The H.323 protocol suite is split up into three main areas.

- **Registration, Admissions, and Status (RAS) Signaling**—It provides pre-call control in H.323 gatekeeper-based networks. The RAS channel is established between endpoints and gatekeepers across an IP network. The RAS channel is

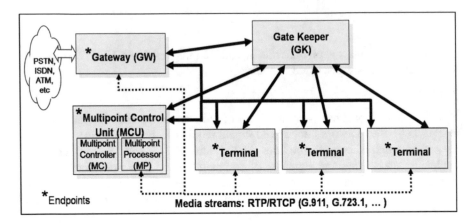

**Fig. 2.1** H.323 elements and their interactions

| Reliable TCP Delivery | | | Unreliable UDP Devlivery | | |
|---|---|---|---|---|---|
| H.245 | H.225 | | Audio/Video Streams | | |
| | Call Control | RAS | RTCP | | RTP |
| TCP | | | UDP | | |
| IP | | | | | |
| Data/Physical Layers | | | | | |

**Fig. 2.2** H.323 protocol layers

opened before any other channel is established and is independent of the call control signaling and media transport channels. The unreliable User Datagram Protocol (UDP) connection carries the RAS messages that perform registration, admissions, bandwidth changes, status, and disengage procedures.

- **Call Control Signaling**—It is used to connect, maintain, and disconnect calls between endpoints. Call control procedures are based on ITU Recommendation H.225, which specifies the use and support of Q.931 signaling messages. A reliable call control channel is created across an IP network on Transmission Control Protocol (TCP) port 1720. This port initiates the Q.931 call control messages between two endpoints for the purpose of connecting, maintaining, and disconnecting calls.

- **Media Control and Transport**—It provides the reliable H.245 channel that carries media control messages. The transport occurs with an unreliable UDP stream. The exchange of capabilities, the opening and closing of logical channels, preference modes, and message control take place over this control channel. H.245 control also enables separate transmit and receive capability exchange as well as function negotiation, such as determining which codec to use.

> The H.323 protocol suite supports call admissions, setup, status, teardown, media streams, and messages in H.323 systems.

## 2.2.3   Call Flow

H.323 creates, maintains, and terminates calls in several phases that are discussed as follows.

### Initializing the call

H.323 initializes the call using SETUP, ALERTING, and CONNECT messages. Terminal A sends a SETUP message containing the number and address along with other information to Terminal B. B then replies with either CALL PROCEEDING, ALERTING, CONNECT OR RELEASE COMPLETE messages. As ALERTING is sent indicating that "the remote phone is ringing," Terminal B picks up the call and sends CONNECT message to Terminal A stating the IP address and port to open H.245 TCP connection, call identifier, protocol identifier, Conference ID, etc.

### Establishing the control channel

It is done in two phases. In the Capability Negotiation Phase, media control and capability exchange messages are sent in the form of Terminal Capability Set by each terminal and acknowledged by the other. In the master–slave determination phase, one of the terminals is chosen as the master for initiating important functions. This is accomplished by exchanging master–slave determination messages.

### Opening media channels

Both Terminals A and B open media channels for voice and possibly video or data. The digitized data is carried in several "logical channels." This is done by exchanging OpenLogicalChannel message that contains UDP address and port, type of RTP payload and capacity to stop sending data during silences.

**Dialogue**

This is the phase where the caller and callee actually communicate using voice and/or video. The media data is sent in RTP packets and the control information is sent via RTCP packets using sender reports (SR) and receiver reports (RR).

> The H.323 call flow comprises of several phases, such as the initiation phase, capability negotiation phase, master–slave determination phase, and finally the dialogue phase.

Gatekeepers are required in scenarios where the caller or the callee may not be using Internet-based phones. The gatekeeper is the most complex component of the H.323 framework and is responsible for most network-based services. There are two call flow methods involving gatekeepers. They are described below.

**Direct endpoint call signaling**

Call signaling messages are sent directly between the two endpoints as shown in Fig. 2.3 [3].

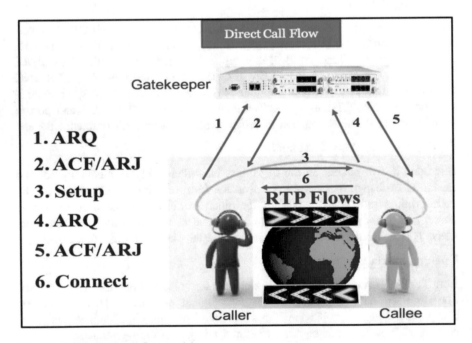

**Fig. 2.3** H.323 direct call flow model

1. Here, both the endpoints send Admission Request (ARQ) messages to the gatekeeper.
2. Based on filtering and other admission control features, the gatekeeper accepts the request by sending Admission Confirm (ACF) messages or rejects the request sending Admission Reject (ARJ) messages.
3. If both endpoints are granted admission, then they exchange setup and connect messages and the call is established.

### Gatekeeper-routed call signaling

Call signaling messages between the endpoints are routed through the gatekeeper [3, 4] as shown in Fig. 2.4. Here, the gatekeeper has centralized control over the entire duration of the call. Unlike the previous scenario, in this case, even the setup and connect messages are routed through the gatekeeper. However, there may be a modification to this model where the RTP flows may be established directly between the two endpoints in order to reduce the latency.

> The gatekeeper is the most complex component of the H.323 framework and is responsible for most network-based services including the direct endpoint call signaling and gatekeeper-routed call signaling flows.

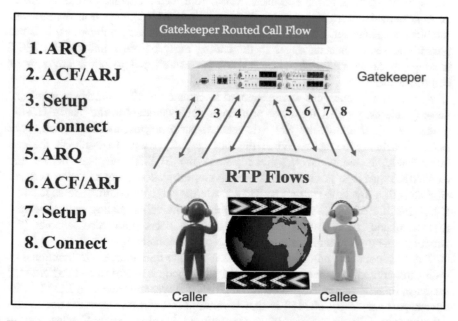

Fig. 2.4   H.323 gatekeeper-routed call flow model

## 2.2.4   Enhancements

There are several issues that hinder the performance of VoIP calls under the supervision of H.323. One of the major weaknesses of the initial version of H.323 is the delay in establishing a call setup. In the simplest scenario, H.323v1 procedures require at least five message communication round-trip times before a call is established. This includes one message for the ARQ/ARJ sequence, another message for SETUP-CONNECT exchange, two messages for H.245 capabilities exchange and master–slave procedure and finally, one message for setup of each logical channel. This situation is further worsened when Q.931 and H.245 channels use TCP connection, which requires additional setup time. As TCP uses three-way handshake for its setup, delays are incurred that may be unacceptable with respect to call setup. There is another drawback of TCP that must be eliminated while implementing it over H.323. During the SETUP-CONNECT message exchange, if the SETUP message is greater than the MTU of the network, TCP breaks the message into segments and sends them one by one and initiates the slow-start procedure. This implies that initially, only a single segment is transmitted. Only on receiving the acknowledgement of the sent message, TCP increases its window size and increases the rate of transmission. Thus, slow-start mechanism further increases the delay.

Another problem of the H.323v1 is its inability to generate messages to users before the media channels are established. In circuit-switched voice networks, "Busy" status is generated by the network at the local exchange during congestion and transmitted back to the end user, even before media channels are established. Auto-generated messages in tele-voting applications also do not require call establishment between two users. However, in H.323-based applications, it is not possible to send voice messages to the calling party before sending CONNECT message, because media channels are not yet established to allow any sort of communication.

A number of techniques were proposed to enhance H.323 in order to minimize these problems. One of the most prominent enhancements is the "early H.245" procedure [5]. It allows the H.245 to start as early as possible in the call, even before it actually connects. When the calling party wants to establish the call immediately, it sends the H.245 address in the SETUP message, before any CONNECT message is received. Alternatively, the called party can also perform similar action by embedding the H.245 address in call control messages like CALL PROCEEDING, ALERTING. This makes the call signaling operations and the associated delays transparent to the end users and also enables the network-generated prompt messages before call establishment.

The "fast connect" procedure is another technique that enables unidirectional or bidirectional media channels to be established immediately after the Q.931 SETUP message, thereby minimizing the call setup delay. It was introduced in H.323v2. Its advantage over "early H.245" is that it minimizes even any post-connect audio delays when calls are connected at the earliest. A calling party that decides to implement "fast connect" includes a new parameter, called fastStart in the SETUP message. fastStart provides a complete list of the all the media channels where the user is prepared to send and receive information and also contains the codecs and other parameters required for call establishment. Once the called party enables "fast connect," it sends back another fastStart in its call control messages (CALL PROCEEDING, ALERTING, etc.) that select from among the media channels offered by the calling party. Eventually, calls are established before any exchange of H.245 messages.

H.245 Tunneling [5] is another option to reduce the time incurred in call setup. In the normal scenario, separate TCP connections are established for Q.931 messages and H.245 messages. The problems due to three-way handshake and slow start in TCP are aggravated further in this approach. Moreover, it also creates problems for gateways or gatekeepers with low limits on TCP connections. This problem is solved by the concept of tunneling where the H.245 messages are encapsulated in the H.245 control element of Q.931 messages by both the calling and called endpoints. This implies that Q.931 channel must be open and active during the entire duration of the call. In the absence of any pending Q.931 message to be sent, H.245 messages can be tunneled in a Q.931 FACILITY message.

A number of techniques were proposed to enhance H.323, such as the early H.245 procedure, "fast connect" procedure and H.245 Tunneling mechanism.

## 2.3   Session Initiation Protocol

There are many applications that require the creation and management of a session, where a session is considered an exchange of data between an association of participants. The SIP is an application-layer control protocol that can establish, modify, and terminate multimedia sessions (conferences) such as Internet telephony calls. SIP can also invite participants to already existing sessions, such as multicast conferences. SIP transparently supports name mapping and redirection services, in addition to personal mobility where users can maintain a single externally visible identifier (SIP URI) regardless of their network location [2].

SIP supports the five facets of establishing and terminating multimedia communications as illustrated in Fig. 2.5.

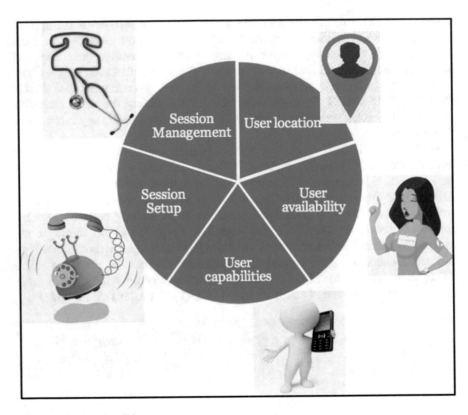

**Fig. 2.5** SIP functionalities

- *User location*: determination of the end system to be used for communication.
- *User availability*: determination of the willingness of the called party to engage in communications.
- *User capabilities*: determination of the media and media parameters to be used.
- *Session setup*: "ringing," establishment of session parameters at called and calling party.
- *Session management*: includes transfer and termination of sessions, modifying session parameters, and invoking services.

SIP is an application-layer control protocol that can establish, modify, and terminate multimedia sessions (conferences) such as Internet telephony calls with simple call flows and messages.

## 2.3.1   SIP Actors

SIP actors refer to all those entities who are involved in the creation, management, and termination of SIP sessions.

### User Agents

#### A.  User agent client (calling party)

A user agent client (UAC) is a logical entity that creates a new request and then uses the client transaction state machinery to send it. The role of UAC lasts only for the duration of that transaction.

#### B.  User agent server (called party)

A user agent server (UAS) is a logical entity that generates a response to a SIP request. The response accepts, rejects, or redirects the request for the entire duration of the transaction.

### Servers

#### A.  Proxy Server

It is an intermediary entity that acts as both a server and a client for the purpose of making requests on behalf of other clients. A proxy server primarily plays the role of routing. Its job is to ensure that a request is passed on to another entity "closer" to the targeted user and is also useful for enforcing various policies. Proxy servers can be stateful or stateless. Stateful proxies keep any state relative to the call and all transactions involved in the call. Stateless proxy keeps no such information. It just chooses the next hop destination for an incoming SIP message using to header information.

#### B.  Redirect Server

A redirect server is a server that generates 3xx responses to requests it receives, directing the client to contact an alternate URI.

#### C.  Registrar

A registrar is a server that accepts register requests and places the information it receives in those requests into the location service for the domain it handles.

## 2.3.2   SIP Structure

SIP is structured as a layered protocol. Its behavior is described in terms of a set of fairly independent processing stages for the purpose of presentation and concise-ness with only a loose coupling between each stage.

The lowest layer of SIP is its syntax and encoding. Its encoding is specified using a Binary Message Form (BNF). The next higher layer is the transport layer. This layer defines how a client takes a request and physically sends it over the network, and how a response is sent by a server and then received by a client. All SIP elements contain a transport layer.

The next higher layer is the transaction layer. A transaction is a request, sent by a client transaction (using the transport layer), to a server, along with all responses to that request sent from the server transaction back to the client. The transaction layer handles application layer retransmissions, matching of responses to requests, and application layer timeouts. User agents and stateful proxies contain a transaction layer. Stateless proxies do not contain a transaction layer [2].

> SIP is structured as a layered protocol.

### 2.3.3   SIP Message Type

SIP is a text-based protocol and uses the ISO 10646 character set in UTF-8 encoding (RFC 3629) [6]. A SIP message is either a request from a client to a server, or a response from a server to a client. The syntax is as follows: generic-message = [start-line] + [message-header] + [CRLF] + [message-body].

#### A.  Requests

SIP requests are distinguished by having a Request-Line for a start-line. A Request-Line contains a method name, a Request-URI, and the protocol version separated by a single space (SP) character. The Request-Line ends with CRLF. The syntax is Method Request-URI SIP-Version.

The various SIP requests are described in Table 2.2.

**Table 2.2**  SIP Requests

| Method | Description |
| --- | --- |
| INVITE | Initiates a call, changes call parameters (re-INVITE) |
| ACK | Confirms a final response for INVITE |
| BYE | Terminates a call |
| CANCEL | Cancels searches and "ringing" |
| OPTIONS | Queries the capabilities of the other side |
| REGISTER | Registers with the location service |

B. **Responses**

SIP responses are distinguished from requests by having a Status-Line as their start-line. A Status-Line consists of the protocol version followed by a numeric Status-Code and its associated textual phrase, with each element separated by a single SP character. The syntax is *SIP-version Status-Code Reason-Phrase*.

SIP responses are of two types:

- Provisional (1xx class)—provisional responses are used by the server to indicate progress, but they do not terminate SIP transactions.
- Final (2xx, 3xx, 4xx, 5xx, 6xx classes)—final responses terminate SIP transactions.
- SIP is a text-based protocol and uses the ISO 10646 character set in UTF-8 encoding (RFC 3629).

## 2.3.4   SIP Call Flows

The different SIP call flows involving the various SIP actors (as described above) are shown.

A. **SIP Call Flow with Proxy Server**

Figure 2.6 describes a SIP call flow with proxy server. The steps are described as under.

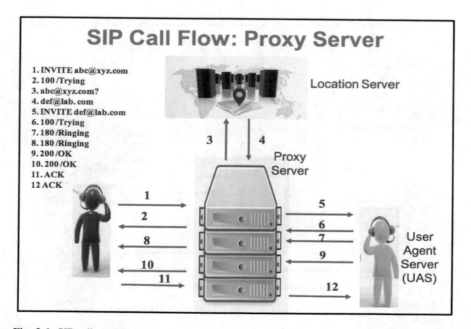

**Fig. 2.6** SIP call proxying

1. The proxy server accepts the INVITE request from the client.
2. The proxy server identifies the location by using the supplied addresses and location services.
3. An INVITE request is issued to the address of the location returned.
4. The called party user agent alerts the user and returns a success indication to the requesting proxy server.
5. An OK (200) response is sent from the proxy server to the calling party.
6. The calling party confirms receipt by issuing an ACK request, which is forwarded by the proxy or sent directly to the called party.

B. **SIP Call Flow with Redirect Server**

1. The redirect server accepts the INVITE request from the calling party and contacts location services with the supplied information.
2. After the user is located, the redirect server returns the address directly to the called party. Unlike the proxy server, the redirect server does not issue an INVITE.
3. The user agent sends an ACK to the redirect server acknowledging the completed transaction.
4. The user agent sends an INVITE request directly to the address returned by the redirect server.
5. The called party provides a success indication (200 OK), and the calling party returns an ACK.

Figure 2.7 describes a SIP call flow with redirect server.

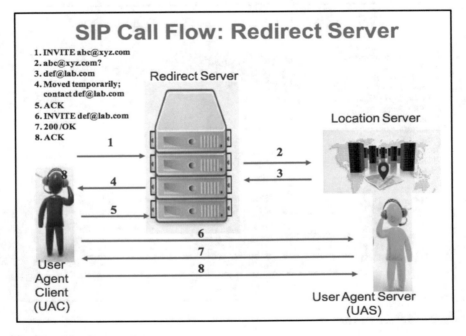

**Fig. 2.7** SIP call redirection

SIP call flows can occur either through SIP proxy server or SIP redirect
server.

## 2.4   Megaco (H.248)

Based on a master–slave approach, this protocol manages call sessions across
Media Gateways (MGs) and is widely used to connect VoIP calls to traditional
PSTN telephony. This protocol as developed by IETF and ITU [7] subsequently
separates the call control logic from the media conversion in a gateway. There are
several MGs in a network that are controlled by a Media Gateway Controller
(MGC). A MGC with its subordinate MGs constitutes a "calling" domain. These
MGCs further communicate among each other via messages using the SIP messaging protocol. The sources and sinks in Megaco are referred to terminations
which may be either physical or ephemeral. Connections are achieved by placing
one or more terminations into a Context (CTX). The relation among these entities is
illustrated in Fig. 2.8.

### 2.4.1   Call Flow Description

The call flow comprises of two parts. In part 1, MG communicates with the MGC
controlling it. In the second part, two MGCs exchange messages among each other
[8]. The overall process can be grouped into several stages as described below.

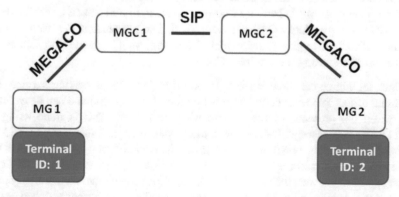

**Fig. 2.8** Megaco entities

A. **Registration**: Before any communication is made possible, every MG registers with its MGC using the Service Channel Restart Command. MGC in return accepts the gateway and evaluates its capabilities. Accordingly, it sends an Audit message to the MG1. After receiving Capabilities message from the MG1, the MGC modifies the terminations with an Embedded Event Descriptor, instructs the MG to detect an off-hook event, plays a dial-tone on off-hook, and detects the digital digits.

B. **Call**: As from Fig. 2.8, Terminal ID 2 goes off-hook and makes a call to Terminal ID 1 via MG1. MGC 2 determines that MG1 is not registered to it, but to MGC 1. Therefore, MGC 2 sends a SIP INVITE message to MGC 1 along with the audited capabilities. Also, it ensures in the INVITE message that the ensuing responses must be reliably delivered to it. MGC 1 compares the received Capabilities from MGC 2 and compares them with the audited ones from MG 1 (that it received during registration). Based on the comparison, it prepares a set of common capabilities and sends it to the MGC 2 using SIP 183 Session Progress Response. MGC 2 acknowledges the receipt using PRACK message.

C. **SDP Negotiation**: MGC 2 specifies the Local SDP as sent by MGC 1 to MG2 by issuing an ADD command. MG2 creates an Ephemeral Termination, selects an SDP (SDP3), and sends a Megaco Local Descriptor message to MGC 2 as a reply to the ADD command, where it informs the callee of its chosen capabilities. MGC 2 transmits this remote SDP to MGC 1 through SIP UPDATE request. MGC 1, in turn, provides MG 1 with the new SDP using ADD command. MG 1 chooses its Ephemeral Termination and Capabilities, returns them to MGC 1, and also plays the Ringtone. MGC 1 forwards this SDP to MGC 2 in a 200 OK Response message to the previously sent UPDATE request. MGC 1 further sends the 180 Ringing Response to MGC 2. MGC 2 modifies its MG 2 with the new SDP and the 180 Ringing Response and finally starts the Ring Back Tone.

D. **Media Flow**: This involves the actual VoIP communication. The callee picks up the phone in response to the Ringing tone. MGC 2 changes the stream code to send and receive and initiates the RTP Media Stream.

E. **Call Termination**: When any one user decides to end the call, it informs the MGC using the Notify OnHook Command. The remote side is informed using the SIP BYE request after which both the MGCs delete the Ephemeral Terminations using the Subtract Command.

Thus, the above call flow is strongly based on the SIP messaging service. One important aspect to be mentioned here is that the Megaco command can be executed only when the presence of the remote party is known. This is done using the INVITE request message. It determines the presence of the other party (MGC 1 in this case). As a result, 11 SIP messages are exchanged between the two parties. One way to reduce this is to use the OPTION request message that results in only eight message passes between the two MGCs. Also, if the reliable message delivery is not enforced, the total number of messages transmitted can be further reduced.

## 2.4.2 Command Format

The command format in Megaco is comprised of the following fields.

- Termination ID
- Local Termination State
- Local Termination Descriptor, Remote Termination Descriptor (description of the media flow in each direction, e.g., IP address of the endpoint, port address, codec, etc.)
- Events Descriptor (lists all the events to be reported (trigger))
- Signal Descriptor (list of signals to be applied at the termination; signals are tones and announcements generated by the MG)

The basic commands in Megaco and their functionalities are listed in Table 2.3.

## 2.4.3 Megaco and PSTN

Based on the call flows and the commands as already described, Megaco is used for IP-based communication and VoIP services between IP networks and PSTN or can be used entirely within IP networks. A basic architectural model for Megaco in PSTN is shown in Fig. 2.9.

Based on a master–slave approach, Megaco protocol manages call sessions across Media Gateways (MGs) and is widely used to connect VoIP calls to traditional PSTN telephony.

**Table 2.3** Megaco commands

| Command | Description |
|---|---|
| Add | Adds a termination to a CTX |
| Modify | Modifies the properties, events, and signals of a termination |
| Subtract | Deletes a termination from CTX |
| Move | Moves a termination from a CTX |
| Audit value | Returns current state of properties, statistics, signals, and events of a termination |
| Audit capabilities | Returns all possible values for termination properties, events, and signals |
| Notify | Used by MGs to notify events to MGCs |
| Service change | Used by MGs to (i) notify MGCs that a group of terminations have been taken out from service or returned to service, (ii) register with MGC upon restart, (iii) during handover |

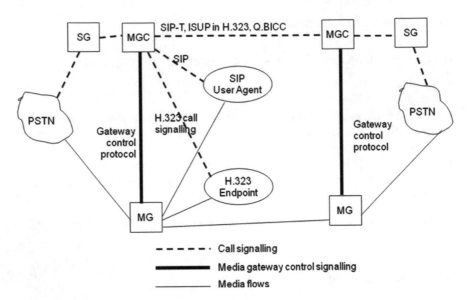

**Fig. 2.9** Operation of Megaco in PSTN via IP networks

## 2.5   Security Issues in H.323 and SIP

Applications require various procedures to ensure that the transmission is secured while the packets are routed through best-effort networks such as Internet. Although personal communications may treat security as an optional requirement, it becomes a major concern when communication is established in enterprise, military, banking and related sectors that treat any loss in the integrity and privacy of data as a serious threat. Apart from the audio and video stream, signaling also requires security, primarily authentication to prevent spoofing of calls, denial-of-service attacks, flooding and generating spam, etc. Therefore, with respect to the call signaling protocols, security includes protecting call setup, call management, and billing services.

Security is handled in H.323 by H.235 [9] that provides authentication of users and protection of the integrity of streams besides securing the data streams. Keys are used for encrypting and decrypting the streams. The exchange of keys between the VoIP endpoints can be done using out-of-band key exchange or by implementing Diffie–Hellman protocol, Oakley Key Determination Protocol, etc. Privacy of the communication is ensured by the application vendors who either incorporate security aspects in their own private protocols or else utilizes Transport Layer Security (TLS) and IP Security Protocol (IPSec) techniques. TLS is based on Secure Socket Layer (SSL) and Private Communication Technology (PCT). H.235 also manages the security during call setup (Q.931), call management (H.245), and gatekeeper registration/admission/status (RAS). The endpoints need to be authenticated and the information in the signaling messages must be preserved. Encryption of the messages is performed to ensure the integrity of information. Authentication, on the other hand, is carried out both by gatekeepers and the clients. Two techniques are available namely symmetric encryption-based authentication, which is a one-direction method requiring no prior contact between the endpoints, and bidirectional subscription-based authentication, requiring some kind of prior contact.

SIP implements HTTP-based security mechanisms during VoIP communications [10]. Caller and callee authentication can be realized with HTTP mechanisms, including basic (clear-text password) and digest (challenge–response) authentication. Keys for media encryption are conveyed using SDP. Improved security mechanisms are found in SIP version 2.1, which defines end-to-end authentication and encryption using either Pretty Good Privacy (PGP) or S/MIME. SIP messages can be encrypted hop-by-hop using IPSec. They can also be transported over a secured transport layer such as TLS. Thus, security mechanisms are enabled by call signaling protocols based on the user requirements and the network conditions, while maintaining the QoS parameters within their threshold limits.

> Apart from the audio and video stream, signaling also requires security, primarily authentication to prevent spoofing of calls, denial-of-service attacks, flooding and generating spams, etc.

## 2.6   H.323 and SIP: Comparison

Both H.323 and SIP have witnessed widespread usage in VoIP telephony. However, there are several factors that must be taken into consideration before selecting the call signaling protocol for a particular communication setup [11]. Hence, a comparative study of these protocols is essential.

### 2.6.1   Addressing

Each physical H.323 entity has one network address, which is dependent on the network. Entities may have several TSAP identifiers, allowing multiplexing of several channels sharing the same network address. Dynamic TSAP identifiers are used for the H.245 control channel and media channels. The call signaling channels and RAS channels use well-known TSAP identifiers. Endpoints may also have one or several alias addresses, representing an endpoint or a conference that an endpoint is hosting. A gatekeeper is needed to resolve all these aliases. The alias can consist of telephone numbers, email addresses, URLs, transport IDs, party number or text identifiers.

SIP, on the other hand, uses an email-like identifier in the form "user@host". The user part is a civil name or a telephone number. The host part is either a domain name or a numeric network address. Email-like names can be mapped by any device on the Internet. In addition to individual persons, an address may specify the first available person from a group of individuals or a whole group.

## 2.6.2   Complexity

H.323 has often received criticism for being too heavy, complex, and inflexible. The complexity can be primarily attributed to the multiple protocol components in H.323. These components are tightly intertwined and cannot be separately used or exchanged. On the contrary, SIP is a noticeably simpler protocol and involves fewer and straightforward steps in establishing and managing calls.

Moreover, H.323 protocol is based on ASN.1 and packet encoding rules (PER) and uses a binary representation, thereby requiring complex code-generators to parse. SIP comprises of text-based messages like in HTTP, and therefore, parsers for SIP can be implemented with text processing languages, such as Perl, Tcl, and Java. It is also easier to manage and debug as well as create manual entries of text-based messages in SIP.

## 2.6.3   Call Setup

Initial call setup incurred higher delays for H.323 as compared to SIP. This is because unlike the simpler implementation of SIP call setup that involves fewer messages, H.323 required setup of the Q.931 and H.245 connections before calls can be established. However, with later versions, procedures such as "early H.245" and "fast connect" minimized the overall call setup delays in H.323.

## 2.6.4   Extensibility

As Internet telephony is continuously under development, it is likely that additional signaling capabilities will be needed in the future with the appearance of newer applications day by day. Subsequently, vendors add new features and support for their equipment. However, this is only possible when extensions could be added to the existing protocols. Both H.323 and SIP support extensibility. H.323 provides *nonstandardParam* fields placed in various locations in the ASN.1 structures. These fields consist of a vendor code and a value that can be provided by the vendor. SIP also allows new headers to be added to its messages. Using the *Require* header, the client can specify headers to be understood by the other endpoint.

The bottom line is:
Want a simpler solution at the cost of reliability??
Use SIP
Want robust and reliable solution at the cost of higher complexity??
Use H.323

In a nutshell, the similarities and differences between the two protocols are illustrated in the underlying chart.

# SIP and H.323 : Comparison

Similarities:

- Use RTP/RTCP for media transport
- Support call routing through proxies/gatekeepers and can be stateless/stateful
- Same set of voice codecs
- Similar call-setup times (3)
- Authentication (H.235 in H.323, HTTP, SSL in SIP)

Differences:

- Encoding (Text vs ASN, PER)
- Protocol structure (Single in SIP, Umbrella standard in H.323)
- Load balancing (Trivial in SIP, Robust in H.323)
- Backward compatibility (more in H.323, less in SIP)
- Complexity (SIP simpler)
- Conferencing (limited for SIP, full support for H.323)
- Reliability (H.323 more reliable)

## 2.7   Current Status of H.323 and SIP

### 2.7.1   H.323

H.323 was first formally defined in the work titled "Visual Telephone Systems And Equipment For Local Area Networks Which Provide A Non-Guaranteed Quality Of Service" [5] way back in 1996. Since then, several versions of H.323 have been proposed till date. Versions 2, 3, and 4 addressed many decontrol, caller-id, gateway decomposition, multiplexed stream transmission, etc. H.323 version 5 aimed to maintain stability in the protocol by introducing only modest additions to the base protocol rather than making sweeping changes. The primary changes to the base H.323 specification are the introduction of the "Assigned Gatekeeper" in version 6 released in 2006 [5]. Enhancements were made in error handling, call clearing procedures, etc. Two new features introduced into the final version 7 of H.323 in 2009 are single transmitter multicast and introduction of MCU for conferencing [5]. In this way, H.323 has gone a long way in ensuring success of VoIP.

## 2.7.2   SIP

There has been continuous monitoring and development with respect to SIP and its related features over the years since 2002 when the definition of SIP was made available in RFC 3261 [2]. RFC 5658 [12], for instance, addresses record-route issues in the SIP. A framework for application interaction in the SIP is mentioned in RFC 5630 [13], whereas session mobility has been further described in RFC 5631 [14]. Call control capabilities in SIP are broadly defined in RFC 5589 [15]. Emergence of 3GPP networks motivates the specification of SIP P-Served-User P-header in RFC 5502 [16]. RFC 5479 [17] describes requirements for a protocol to negotiate a security context for SIP-signaled Secure RTP (SRTP) media. Essential requirements for real-time Text-over-IP (ToIP) and a framework for implementation of all required functions based on SIP and RTP are proposed in RFC 5194 [18]. Thus, SIP protocol has been updated continuously to keep pace with the changing scenarios.

## 2.8   Summary

Increasing importance is being meted out to select appropriate call signaling protocol before deploying VoIP in practical networks. This is because efficiency of the protocol plays a critical factor in ensuring high quality of VoIP communication. Both SIP and H.323 have their own advantages and disadvantages and are utilized by the vendors depending on their specific requirements. Technological advancements have also made it possible to integrate both these protocols using gateways and various gateway control protocols (such as Megaco). Accordingly, metrics must be defined and established that will judge the performance efficiency of these protocols and also allow monitoring of their activities. As innovative applications continue to utilize the services of VoIP, call signaling protocols undergo continuous enhancements in terms of their protocol structure and functional specifications.

## References

1. Recommendation H.323, H.323: packet-based multimedia communications systems by ITU-T. Article No. E35738 (2010)
2. J. Rosenberg et al., *SIP: Session Initiation Protocol.* IETF RFC 3261, June 2002
3. CISCO, *Understanding H.323 Gatekeepers.* Document ID 5244, Sept 2014
4. K. Wallace, *Implementing Cisco Unified Communications Voice Over IP and QoS (Cvoice) Foundation Learning Guide* (Cisco Press, 2011)
5. H.323 FORUM. Available: http://www.h.323forum.org
6. F. Yergeau, *UTF-8, a Transformation Format of ISO 10646.* RFC 3629, Nov 2003
7. Recommendation H.248.1, *H.248.1: Gateway Control Protocol: Version 3.* ITU-T. Article No. E38602 (2013)

8. T. Taylor, Megaco/H.248: a new standard for media gateway control. IEEE Commun. Mag. **38**(10), 124–132 (2002). https://doi.org/10.1109/35.874979
9. Recommendation H.235, Amendment 1, H.235: security and encryption for H-series (H.323 and other H.245-based) multimedia terminals by ITU-T. Article No. E25796 (2003)
10. J. Arkko et al., *Security Mechanism Agreement for the Session Initiation Protocol (SIP)*. RFC 3329, Jan 2003
11. H. Schulzrinne, J. Rosenberg, A comparison of SIP and H. 323 for internet telephony, in *Proceedings of International Workshop on Network and Operating System Support for Digital Audio and Video (NOSSDAV)*, pp 83–86 (1998)
12. T. Fromet et al., *Addressing Record-Route Issues in the Session Initiation Protocol (SIP)*. IETF RFC 5658, Oct 2009
13. F. Audet, *The Use of the SIPS URI Scheme in the Session Initiation Protocol (SIP)*. IETF RFC 5630, Oct 2009
14. R. Shacham et al., *Session Initiation Protocol (SIP) Session Mobility*. IETF RFC 5631, Oct 2009
15. R. Sparks et al., *Session Initiation Protocol (SIP) Call Control-Transfer*. IETF RFC 5589, June 2009
16. J. van Elburg, *The SIP P-served-User Private-Header (P-Header) for the 3GPP IP Multimedia (IM) Core Network (CN) Subsystem*. IETF RFC 5502, Apr 2009
17. D. Wing, S. Fries, H. Tschofenig, F. Audet, *Requirements and Analysis of Media Security Management Protocols*. IETF RFC 5479, Apr 2009
18. A. van Wijk, G. Gybels, *Framework for Real-Time Text Over IP Using the Session Initiation Protocol (SIP)*. IETF RFC 5194, June, 2008

# Chapter 3
# Quality of Service Management—Design Issues

## 3.1 Introduction

Every application is bounded by certain constraints that must be satisfied for its efficient functioning. In other words, the "Quality of Service" provided by the application increases, once all the relevant constraints are properly addressed. Quality of Service (QoS) is, hence, defined as the ability to provide different priority to different applications, users, or data flows, or to guarantee a certain level of performance to a data flow. With respect to VoIP, QoS refers to maintaining a standard quality of voice communication with respect to the end users.

QoS guarantees are important if the network capacity is insufficient and assumes significance, especially for real-time streaming multimedia applications such as VoIP, online games, and IP-TV, since these often require fixed bit rate and are delay sensitive. Networks where the capacity is a limited resource, for example in cellular data communication, [1] also demand QoS guarantees. Moreover, in modern intelligent networks like cognitive radio networks, QoS guarantees must be classified with respect to different sets of users and varying applications deployed.

QoS guarantees for every application are dependent on a set of application-specific parameters that must be controlled within certain limits. Every such parameter is again dependent on a set of factors that should be identified and analyzed accordingly. With respect to VoIP, it has been noticed that several attributes ascertain the QoS parameters like delay, jitter, and packet loss. Primarily, the call signaling protocol used, the networking environment being implemented, the user capacity supported by the system and related issues affect the performance of VoIP communication [2].

> QoS guarantees are important if the network capacity is insufficient and assumes significance especially for real-time streaming multimedia

© Springer International Publishing AG, part of Springer Nature 2019

T. Chakraborty et al., *VoIP Technology: Applications and Challenges*,
Springer Series in Wireless Technology, https://doi.org/10.1007/978-3-319-95594-0_3

applications such as VoIP, online games, and IP-TV, since these often require
fixed bit rate and are delay sensitive.

## 3.2  QoS as Service

A network or protocol that supports QoS enters into a traffic contract with the
application software and reserves capacity in the network nodes, for example,
during a session establishment phase. The protocol measures the performance level
achieved during the session and controls the traffic flows and scheduling priorities
besides implementing call admission control algorithms. The protocol follows
certain QoS policies that state which client, application, and schedule must receive a
particular service. Accordingly, QoS as service can be classified as differentiated
service and integrated service [3].

A.  Differentiated Service

This is the first type of outbound bandwidth policy that is implemented on the
server. Differentiated service [4] divides the overall network traffic into classes and
implements separate QoS policies to handle the different classes. The most common
classes are defined using client IP addresses, application ports, server type, protocol, local IP address, and schedule. Equal treatment is meted out to all traffic
belonging to a same class.

After traffic is classified, differentiated service handles the traffic based on a
per-hop behavior (PHB). The server uses bits in the IP header to identify an IP
packet's level of service. Routers and switches allocate their resources based on the
PHB information in the IP header's type of service octet (TOS) field. For a packet
to retain the service requested, every network node must be differentiated service
(DiffServ)-aware and should be capable of enforcing per-hop behaviors. To enforce
PHB treatment, the network node should be able to use queue scheduling and
outbound priority management. It must be noted that in the absence of Diffserv
policy at a certain intermediate network element, the packet is still handled, but it
may experience unexpected delay.

Traffic conditioners as denoted by Fig. 3.1 form an integral part of Diffserv
aware QoS policies. If the network equipment has all the traffic conditioners, then it
is considered to be completely DiffServ-aware.

Differentiated service divides the overall network traffic into classes and
implements separate QoS policies to handle the different classes.

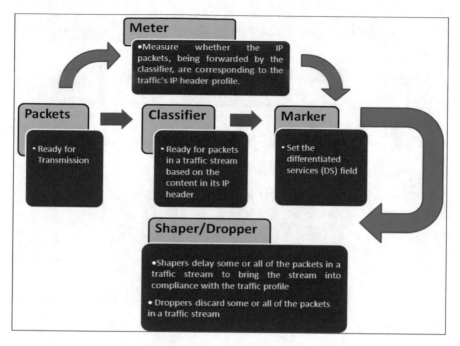

**Fig. 3.1** Traffic conditioners for DiffServ QoS policy

## B.  Integrated Service

Integrated service (IntServ) [5] is another type of outbound bandwidth policy that deals with traffic delivery time and assigns particular traffic special handling instructions. The basic principle of integrated service is resource reservation following a set of rules before data transmission. The routers are signaled before data transfer and the network agrees to and manages (end-to-end) data transfer based on the designed set of rules.

It is basically an admission control policy. The client requests for bandwidth reservation based on the application. If all the routers in the path agree to the requirements coming from the requesting client, the request gets to the server and IntServ policy is applied. If this request falls within the limits defined by the IntServ policy, the QoS server grants permission for the Resource Reservation Protocol (RSVP)-based connection and will then reserve the bandwidth for the application.

Every intermediate node handling the traffic must be able to implement the RSVP protocol. The routers implement it using three traffic control functions, namely packet scheduler, packet classifier, and admission control. Therefore, the crucial phase in implementing integrated services policies is being able to control and predict the resources in the network. The logical functioning of Intserv policy is depicted in Fig. 3.2.

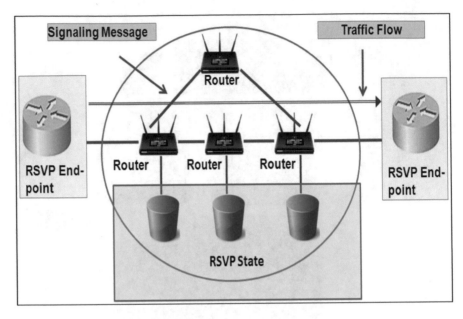

**Fig. 3.2** Logical flow for IntServ QoS policy

IntServ policies can be of two types, namely controlled load and guaranteed service. Controlled load service is suitable for applications that are highly sensitive to congested networks, such as real-time applications. Traffic will be provided with service resembling normal traffic in a network under light conditions. Guaranteed service, on the other hand, assures that packets will arrive within a designated delivery time. Applications that need guaranteed service include video and audio broadcasting systems that use streaming technologies.

> The basic principle of Integrated Service is resource reservation following a set of rules before data transmission.

## 3.3   QoS Framework

QoS framework is constructed in accordance with the QoS service categories as discussed above. The design of a generalized QoS framework is motivated by several QoS principles [6]. Integration principle, for example, states that QoS must be configurable, predictable, and maintainable. Separation principle, on the other hand, states that the tasks of data transfer, control, and management should be separated in architectural QoS frameworks. Transparency principle states that applications must be protected from the complexity of underlying QoS specification and QoS management. Similarly, other principles like multiple timescales principle and performance principle focus on building QoS framework that delivers in a timely and efficient fashion.

The designed QoS framework is enriched with QoS implementation mechanisms which are based on a set of QoS specifications [6]. They include flow performance specification, level of service, QoS management policy, cost of service, and flow synchronization specification. Accordingly, a set of QoS configurable interfaces must be defined, providing a framework for the integration of QoS control and management mechanisms. The generalized QoS framework is further guided by various control mechanisms that provide real-time traffic control of flows based on requested levels of QoS established during the QoS provision phase. The most notable among the control mechanisms is flow shaping that regulates traffic flow based on flow specifications as requested by the user. Flow scheduling is another such control mechanism that supervises the management and forwarding of traffic flows by scheduling them through network systems based on scheduling policies adopted by the user. Another significant control mechanism is flow synchronization which is required to control the event ordering and precise timings of multimedia interaction to avoid unexpected and undesirable network performance.

Finally, various QoS policies ensure that the principles and control mechanisms are adhered to for successful development of the QoS framework. It includes QoS monitoring (that monitors the QoS issues in the network), QoS maintenance (that compares the monitored QoS against the expected performance and then exerts tuning operations), QoS degradation (that issues a QoS indication to the user when performance degradation occurs despite QoS implementation), and QoS scalability (that performs both QoS filtering and QoS adaptation). The overall QoS framework is depicted in Fig. 3.3.

> QoS framework is enriched with QoS implementation mechanisms which are based on a set of QoS specifications.

## 3.4   QoS Parameters for VoIP Applications

A.  Delay

VoIP delay or latency is defined as the total time required by speech to reach the listener's ear from the speaker's mouth. The permissible values of delay that can be tolerated by VoIP applications are standardized by The International Telecommunication Union (ITU) in Recommendation G.114 [7]. This recommendation defines three bands of one-way delay as shown in Table 3.1

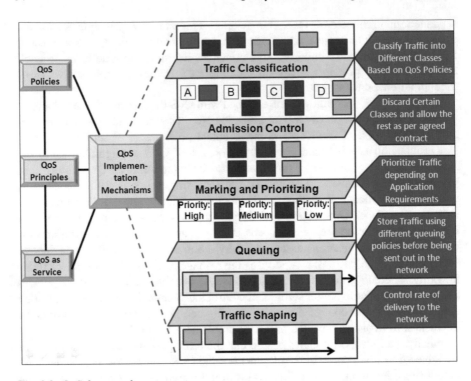

**Fig. 3.3** QoS framework

**Table 3.1** Delay specifications

| Range (ms) | Description |
|---|---|
| 0–150 | Acceptable for most user applications |
| 150–400 | Acceptable provided that administrators are aware of the transmission time and the impact it has on the transmission quality of user applications |
| Above 400 | Unacceptable for general network planning purposes. However, it is recognized that in some exceptional cases, this limit is exceeded |

There are two distinct types of delay with respect components add directly to the overall delay on the connection. Variable delay components not only add to the delay but also introduce "variable delay" or jitter. The components of delay can be categorized as follows.

1. Propagation delay—It is caused by the speed of light in fiber or copper-based networks and is dependent on the distance between the source and the receiver. Speed of light in vacuum is 186,000 miles per second, and electrons travel through copper or fiber at approximately 125,000 miles per second. A fiber network stretching halfway around the world (13,000 miles) induces a one-way

delay of about 70 ms. In order to estimate propagation delay, a popular estimate of 10 ms/mile or 6 ms/km (G.114) is widely used. However, intermediate multiplexing equipment, backhauling, microwave links, and other factors found in carrier networks create many exceptions.

2. Handling delay—Handling delay or processing delay defines many different causes of delay (actual packetization, compression, and packet switching) and is caused by devices that forward the frame through the network. Coder delay is the time taken by the digital signal processor (DSP) to compress a block of PCM samples and varies with the voice coder used and processor speed. This compression algorithm relies on known voice characteristics to process sample blocks and, in this process, induces algorithmic delay. This encoded/compressed speech is filled in the packet payload that results in packetization delay. This delay is a function of the sample block size required by the vocoder and the number of blocks placed in a single frame. The time necessary to move the actual packet to the output queue for transmission results in packet switching delay.

3. Buffering delay—Every network element including routers, access points, callers and callees use buffers to implement the "store and forward" policy in packet switched VoIP networks. When the voice packets are stored in the buffer because of link congestion, buffering delay is introduced. This delay occurs when more packets are sent out that the interface can handle at a given interval. Queuing delay, hence, is a variable delay and is dependent on the state of the queue and network characteristics. Buffers are also used to remove the variable component of delay namely, jitter. Such de-jitter buffers implement various algorithms based on fixed and sliding window concepts and hold the voice samples for a certain time before playing them out. This holding time results in de-jitter delay.

4. Serialization delay—Serialization delay is the amount of time it takes to actually place a bit or byte onto an interface. It is the fixed delay component required to clock a voice or data frame onto the network interface. At high line speed and small frame size, its influence on delay is relatively minimal.

Fixed delay components add directly to the overall delay on the connection. Variable delay components not only add to the delay but also introduce "variable delay" or jitter.

The various delay components involved in VoIP communication are illustrated in Fig. 3.4.

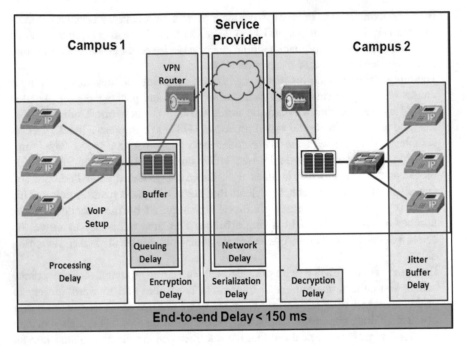

**Fig. 3.4** Delay components in VoIP system

### B. Jitter

Jitter is the variation of packet inter-arrival time that exists only in packet-based networks. The difference between when the packet is expected and when it is actually received is jitter. Mathematically, Eqs.(1) and (2) represent the delay incurred and jitter $J$ respectively [8].

$$D(i,j) = (R_j - R_i) - (S_j - S_i) = (R_j - S_j) - (R_i - S_i)$$
$$J = J + (|D(i-1,i)| - J) \div 16$$

where $S_i$ is the RTP timestamp from packet $i$, $R_i$ is the time of arrival in RTP timestamp units for packet $i$ and $D(i, j)$ is the delay for two packets $i$ and $j$.

Jitter results due to out-of-order delivery of packets at the receiving end. While the sender in VoIP transmission is expected to reliably transmit voice packets at a regular interval based on codec characteristics, the packets may be delayed during transmission and arrive at different time periods, thereby resulting in jitter at the receiving end. Average one-way jitter must be kept at less than 30 ms. While delay can disrupt the overall real-time communication, the presence of considerable jitter degrades the receiving voice quality.

Jitter is minimized using jitter buffer that conceals the inter-arrival packet delay variation and transforms the variable delay into fixed delay component. Recommended practice is to count the number of packets that arrive late and create a ratio of these packets to the number of packets that are successfully processed. This ratio is used to configure the jitter buffer size and accordingly, the playout time for the initial packets in the queue is decided.

> The difference between when the packet is expected and when it is actually received is jitter which can be minimized using jitter buffer algorithms.

### C. Packet Loss

Packet loss is not tolerable in applications where user satisfaction is the primary constraint. VoIP is such an application where packet loss of more than 5% causes degradation in call quality. Packet loss is also used as a metric by many data protocols to know the condition of the network and thereby reduce the number of packets they are sending.

While voice is fed on data networks in the form of VoIP packets, it is essential to successfully transport voice in a reliable and timely manner. Moreover, it is useful to implement Packet Loss Concealment (PLC) [9] mechanisms to make the voice somewhat resistant to periodic packet loss. Cisco implements one such mechanism where whenever a voice packet is not received when expected, it is assumed to be lost and the last packet received is replayed. With G.729 codec, it implies that the packet loss is only 20 ms of speech. Hence, the average listener does not notice the difference in voice quality.

> VoIP is such an application where packet loss of more than 5% causes degradation in call quality.

### D. Mean Opinion Score

Mean Opinion Score gives a numerical indication of the perceived quality of the media received after being transmitted and eventually compressed using codecs [10]. It is calculated in two ways: subjectively and objectively. In the subjective voice testing, a group of listeners is given a sample of speech material and based on its quality, they give it a rating of 1 (bad) to 5 (excellent). The scores are then averaged to get the MOS. When conducting MOS test, there are certain phrases that are recommended by the ITU-T for use [11].

MOS can also be obtained through implementation of software tools that carry out automated MOS testing in a VoIP deployment. Even though human perceptions are not taken into account in such scenarios, the advantage of having such tools is

**Table 3.2** MOS values and their specifications

| Values | Description |
| --- | --- |
| 5 | Perfect. Like face-to-face conversation or radio reception |
| 4 | Fair. Imperfections can be perceived, but sound still clear. This is (supposedly) the range for cell phones |
| 3 | Annoying |
| 2 | Very annoying. Nearly impossible to communicate |
| 1 | Impossible to communicate |

that they consider all the network dependency conditions that could influence voice quality. Some commonly used software includes AppareNet Voice, Brix VoIP Measurement Suite, NetAlly, PsyVoIP, VQManager, and VQmon/EP.

MOS range is listed in Table 3.2. It must be noted that MOS values need not be always expressed in whole numbers. For instance, a value of 4.0–4.5 is referred to as toll-quality and is an indication of a very high-quality call.

Apart from evaluating the QoS of a particular VoIP call, MOS tests are also conducted to compare how well a particular codec works under varying circumstances that include different background noise levels, multiple encodes and decodes. Further, MOS can be used to compare between different VoIP services and providers.

> Mean Opinion Score gives a numerical indication of the perceived quality of the media received after being transmitted and eventually compressed using codecs and can be calculated using both subjective and objective means.

### E.  R-Factor

The R-factor uses a formula to take into account both user perceptions and the cumulative effect of equipment impairments to arrive at a numeric expression of voice quality. It is governed by the E-model as defined in ITU-T Rec. G.107 [12]. The E-model takes into account a wide range of telephony-band impairments, in particular the impairment due to low bit-rate coding devices and one-way delay, as well as the "classical" telephony impairments of loss, noise and echo and produces a scalar quality rating value known as the "Transmission Rating Factor" or R-factor.

The relation between the different impairment factors and R-Factor (as denoted by $R$) is given by the following equation [12].

$$R = R_o - I_s - I_d - I_{e,\text{eff}} + A \tag{3.1}$$

where $R_o$ is the basic signal-to-noise ratio, $I_s$ represents all impairments that occur more or less simultaneously with the voice signal like too loud speech level, non-optimum sidetone, quantization noise, etc., $I_d$ sums all impairments due to delay and echo effects, $I_{e,\text{eff}}$ is an "effective equipment impairment factor," which represents impairments caused by low bit-rate codecs, term $A$ is an "advantage factor," which represents "advantage of access" for certain systems relative to conventional systems, trading voice quality for convenience.

> R-factor is governed by the E-model and takes into account both user perceptions and the cumulative effect of equipment impairments to arrive at a numeric expression of voice quality.

The range of R-factor values with respect to VoIP transmission is enlisted in Table 3.3.

## F. Echo

Echo is an annoying effect that a user experiences as he hears his own voice back after some milliseconds during an ongoing telephonic conversation. It can range from being slightly annoying to unbearable, making conversation unintelligible. The amount of time after which echo is heard varies depending on various factors [13].

Echo can lead to two major drawbacks. It can be loud, and it can be long. The louder and longer the echo, the more annoying it becomes. With respect to VoIP, whenever the delay is high, noticeable echo can be recorded which can adversely affect the overall user satisfaction. To mitigate the effect of echo in today's packet-based networks, echo cancelers are being built into low bit-rate codecs and operated on each Digital Signal Processor (DSP).

## G. Out-of-Order Delivery

Out-of-order delivery of packets is common in every packet switched networks as different packets may take different routes, each resulting in a different delay. As a result, the packets arrive in a different order than they were sent. However, real-time constraints of VoIP make it mandatory that special additional protocols must be in

**Table 3.3** R-factor values

| Range of E-model rating $R$ | User satisfaction level |
| --- | --- |
| $90 > R < 100$ | Best—very satisfied |
| $80 > R < 90$ | High—satisfied |
| $70 > R < 80$ | Medium—some users dissatisfied |
| $60 > R < 70$ | Low—many users dissatisfied |
| $50 > R < 60$ | Poor—nearly all users dissatisfied |

place for rearranging out-of-order packets to an isochronous state once they reach their destination.

The sequence number in the RTP packets can be analyzed to check whether an out-of-order packet has arrived or not. If an out-of-order packet has arrived and the packet with the previous sequence number has not been transmitted, then the packets are rearranged. Otherwise, the packet is discarded upon arrival. Thus, out-of-order packet delivery induces initial time-out delay, jitter (as different packets arrive with different delays) and loss (due to packets being discarded).

## 3.5   QoS Implementation Policies

VoIP flows must always be prioritized with respect to data flows so that QoS metrics can be guaranteed to remain within threshold limits. This can be implemented either at the network level or at the packet level. From the network point of view, Class of Service (CoS) [14], enables a network administrator to group different packet flows, each having distinct latency and bandwidth requirements. Priority can also be assigned at the packet level using the Type of Service (ToS) field in the IP header. Based on TOS values, priority may be assigned for that packet to be placed in the outgoing queue or to be allowed to take a route with minimum latency and/or maximum reliability.

After assigning priorities to VoIP traffic, appropriate QoS implementation mechanisms must be applied to ensure timely and reliable packet delivery. The primary QoS implementation mechanisms that are currently being deployed are described as follows.

A.  Additional Bandwidth

Additional Bandwidth is achieved by implementing Compressed RTP (cRTP) [15] where the 40-byte IP/RTP/UDP header is compressed to 2–4 bytes (2 bytes when no UDP checksums are sent, and 4 bytes when checksums are sent) as shown in Fig. 3.5. cRTP is used on a link-by-link basis to avoid the unnecessary consumption of available bandwidth.

However, the primary disadvantage with cRTP is that it was designed for reliable point to point links with short delays. Hence, it does not perform well over links suffering from high packet loss rate and high delays. In this regard, Compressed RTP (cRTP) has further been enhanced with RObust Header Compression (ROHC) and RObust Checksum-based header COmpression (ROCCO) techniques.

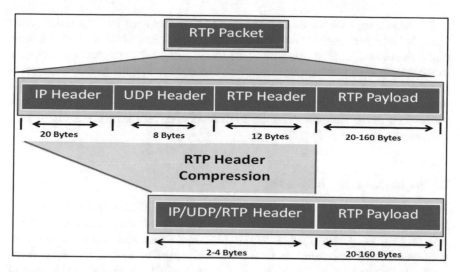

**Fig. 3.5**  Compressed RTP mechanism

### B.  Queuing Policy

Queuing policies play a significant role in maintaining the QoS of VoIP applications. There are different types of queuing policies [16]. First In First Out (FIFO) queuing, for example, places all packets in one queue and transmits them as bandwidth becomes available. Weighted fair queuing (WFQ), on the other hand, uses multiple queues to separate flows and assigns equal amounts of bandwidth to each flow. Custom queuing (CQ) allows users to specify a percentage of available bandwidth to a particular protocol. Priority queuing (PQ) enables the network administrator to configure four traffic priorities namely, high, normal, medium, and low and accordingly, assign each priority to packets arriving from a certain application.

VoIP requires priority but at the same time, data applications must not be starved and they also require some bandwidth guarantees. The principle is that one can use any queuing mechanism depending on several factors like total number of parallel ongoing VoIP transmissions, overall number of data and VoIP applications in the system, the presence of bandwidth reservation protocols, etc.

### C.  Shaping Traffic Flows and Policing

It is necessary to regulate the amount of traffic VoIP application is allowed to send across IP networks in order to maintain overall QoS and allow fairness to every VoIP flow. Rate-limiting tools are implemented in this aspect to drop traffic based upon various traffic policing principles [17]. However, considering a maximum 5% loss allowed in VoIP transmissions, such packet dropping policies are rarely applied to VoIP applications.

Traffic shaping tools are more frequently applied, that generally stores the excess traffic in buffer and waits for the next open interval to transmit the data. In this way, traffic shaper synchronizes the flow of the traffic going out of the network interface to the speed of the remote, target interface and ensures that the traffic conforms to policies contracted for it. While traffic policing refers to dropping any excess packet that arrives after the maximum data rate is obtained, traffic shaping stores such packet in a queue and schedules for its transmission at a later stage. Both these policies are depicted in Fig. 3.6.

### D.  Fragmentation and Interleaving

Fragmentation is required when VoIP packets incur serialization delay as they wait for packet from other applications to be transmitted. In the worst scenario, if a large data packet having size equal to MTU (1500 bytes for serial and 4470 bytes for high-speed serial interfaces) is in the network interface and is currently being serviced, then even a high priority VoIP packet will have to wait for a time of 187.5 ms considering a link speed of 64 kbps. Such delays are inacceptable and hence, a mechanism is needed to ensure that the size of one transmission unit is less than 10 ms. Packets having more than 10-ms serialization delay must be fragmented into 10-ms chunks.

Simple fragmentation is inefficient as even in such case, VoIP packet has to wait till all the fragmented data packets have been serviced. Hence, fragmentation must be combined with interleaving process where VoIP packets are put in between data packets to ensure that serialization delay is kept at a minimum [3]. The process of fragmentation and interleaving is shown in Fig. 3.7.

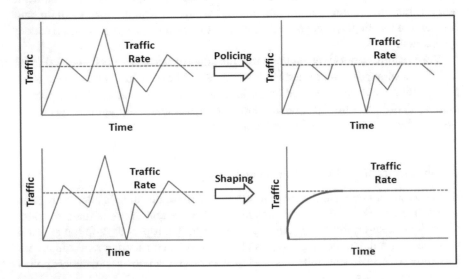

**Fig. 3.6**  Traffic policing and traffic shaping

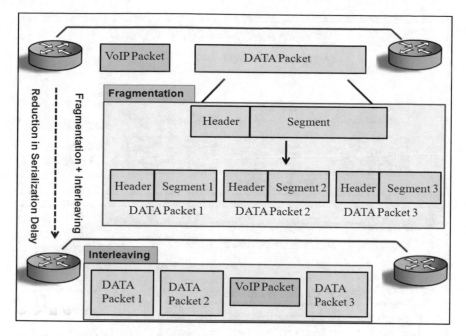

**Fig. 3.7** Fragmentation and interleaving

### E. Jitter Buffer

VoIP systems must ensure that packets are played out at the receiver end in a timely manner and in the correct order. Playout buffers [18] are deployed at the VoIP terminals for this cause. Their primary function is to collect and store packets and play them at an appropriate time. Whenever packets experience variable delay (or jitter), playout buffers can play packets that have arrived later, to ensure the completeness of speech and overall quality of VoIP transmission. Thus, playout buffers help to reduce jitter in VoIP calls and are also called jitter buffers. Playout time is the most important parameter that guides successful implementation of jitter buffers. The playout time can either be fixed or adaptive and depends on the network conditions and user requirements. Fixed playout time introduces constant delay and is relatively simple to implement. Adaptive playout time is controlled by jitter buffer algorithm that calculates the silence time periods in between VoIP talkspurts and accordingly, speeds up or slows down the playout time of the packets waiting at the queue. The basic mechanism of jitter buffer is shown in Fig. 3.8.

Currently, adaptive jitter buffers [19] have gained significance in the wake of increasing bandwidth congestion those results in unpredictable network behavior. These buffers implement different algorithms that vary in their estimation/ calculation of the delay incurred to play a packet. The exponential average algorithm, for example, utilizes the mean and variance to estimate the playout delay of

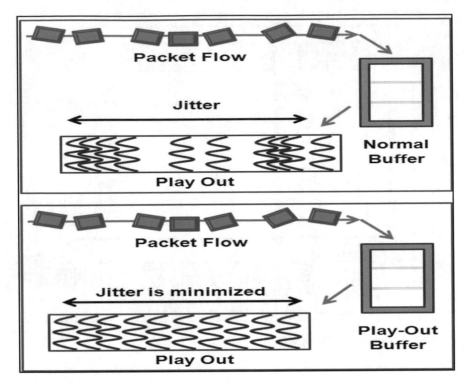

**Fig. 3.8** Playout buffer

the $i$th packet using a weighted () mechanism. This algorithm is further modified to include another weight () that allows the TCP retransmission timer estimate to adapt more quickly to short burst of packets incurring long delays and is termed as fast exponential average algorithm. The minimum delay algorithm minimizes the delay by using the minimum delay of all packets received in the current talkspurt to predict the next talkspurt's playout delay. Analysis of delay patterns in VoIP traffic points to the existence of spikes that are reduced by spike detection algorithm. Recent focus is on developing "window" based jitter buffer algorithms to maximize the perceived user quality (MOS).

In a nutshell, different QoS policies are being implemented both from the systems perspective as well as individual VoIP applications, so as to increase the overall call quality. This assumes special significance when the low-cost VoIP calls are opened for the masses, where maintaining the customer satisfaction proves immensely crucial for the successful deployment of VoIP technology.

After assigning priorities to VoIP traffic, appropriate QoS implementation mechanisms must be applied to ensure timely and reliable packet delivery.

## 3.6   Quality of Experience

Quality of Experience (QoE) is a measure of the overall level of customer satisfaction with a particular vendor. This paradigm can be applied to any consumer-related business or service and is often used in information technology and consumer electronics. Although QoE and QoS are similar in their definitions, both are unique and should be separately considered during performance evaluation of the proposed system. Precisely, QoE expresses the satisfaction of the end customer, both subjectively and objectively. It is therefore user-dependent and depends on the psychological aspects of the customers. Major factors affecting the QoE include cost, reliability, efficiency, privacy, security, interface, user-friendliness and user confidence. At the same time, external factors include the user's terminal hardware (e.g., landline or cellular), the working environment (e.g., fixed or mobile) and the importance of the underlying applications (e.g., text or voice/video).

> QoE expresses the satisfaction of the end customer, both subjectively and objectively.

ETSI (European Telecommunications Standards Institute) has clearly defined the distinctions between the QoS and QoE with respect to wireless communication [20] and the same holds true for VoIP applications as well. Accordingly, QoE for real-time communication can be defined as the overall acceptability of an application or service, as perceived subjectively by the end-user. This definition has two significant aspects.

(i)  QoE encompasses the complete effects of the end-to-end system (including client, terminal, network, services, infrastructure.)
(ii)  Acceptability in terms of user expectations and behavior

For the success of the communication purposes, the overall quality must be judged from three perspectives as stated below:

1. Technically centered Quality of Service (QoS)
2. User-perceived QoS (also known as QoP: Quality of Perception)
3. Quality of Experience (QoE)

These three qualities are related to each other and must be tuned in accordance with the overall objective of increasing the "Quality" of the communication. The relation is shown in Fig. 3.9.

With respect to VoIP, QoE can be easily illustrated with an example. "If a phone call of 30 min duration falters at the last minute when most likely people exchange greetings, the overall QoE of the communication will fall rapidly, even though the QoS is maintained satisfactorily." Although mostly subjective, the voice QoE for VoIP call can be technically expressed as follows [21].

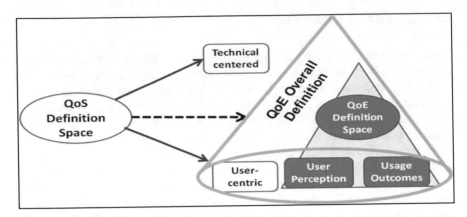

**Fig. 3.9** Definition domain for quality of experience (QoE)

Voice QoE = (Quality of the Network Delivery Stream + Quality of encoding components to the Network + Quality of decode from the network + Human Factors) per unit time.

QoE for real-time communication can be defined as the overall acceptability of an application or service, as perceived subjectively by the end-user.

## 3.7  Current Approach

Networks vary with respect to latency, packet loss rate, bit error rate, peak congestion hour, etc. As congestion increases coupled with other network impairments, even the existing QoS implementation mechanisms seem to deteriorate. Therefore, current QoS policies focus on being adaptive to the changing scenarios. Active queue management [22], for example, has replaced the static buffers with tail drop mechanisms to adaptively deal with providing fairness to each traffic flow while maintaining low packet loss rate. Further, adaptive switching of codecs [23] has gained importance in terms of maintaining VoIP calls even as congestion increases. Focus is also on providing cross-layer optimizations in modern-day networks to maintain QoS. Internet congestion control can be viewed as distributed primal or dual solutions to a convex optimization problem that maximizes the aggregate system performance (or utility) [24]. To cite an instance, a cross-layer optimization strategy is proposed in [25] that optimizes physical layer, data link layer, and application layer using an objective function to maximize user satisfaction with respect to video streaming over wireless networks. Presently, cross-layer optimizations help to realize the QoS through implementation mechanisms as stated above. SWAN architecture, for instance, has modules in different network layers that interact with each other to realize QoS and is described in Fig. 3.10.

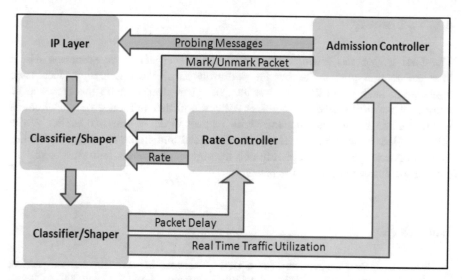

**Fig. 3.10** SWAN basic architecture

Wireless networks currently dominate the network paradigm via the usage of IEEE 802.11 wireless communication standard and recent developments in the fields of wireless ad hoc communication protocols. However, it introduces additional problems, especially those strongly attached to the physical environment namely, effects of data transmission to and from moving stations, type of surrounding building infrastructure and other physical obstruction, external interference phenomena, security issues, etc. Decentralized nature of ad hoc networks further adds to the problems [26].

A cross-layer optimization strategy can be proposed that optimizes physical layer, data link layer, and application layer using an objective function to maximize user satisfaction with respect to video streaming over wireless networks.

Therefore, QoS maintenance in such domain requires efficient power regulation in the physical layer. The data link level QoS can be realized through three ways, namely, (1) MAC protocols that are QoS centralized, (2) Modifications in the existing MAC protocols that belong to the IEEE 802.11 WLAN standard, and (3) Generic MAC protocols that are compatible with these MAC protocols [26]. In addition, QoS implementation in the network layer requires deploying such QoS routing protocols that should efficiently exploit the network resources and provide smooth performance degradation to delay-sensitive traffic, for instance, as done in dynamic source routing and distributed QoS routing in mobile ad hoc networks.

## 3.8   Summary

Maintaining QoS has assumed utmost importance especially in scenarios where
user satisfaction guides the overall performance of the application. Real-time
applications such as VoIP require strict QoS guarantees even in the presence of
congested network scenarios. Wireless networks further introduce new challenges
in the domain of QoS maintenance with respect to delay-sensitive traffic. While
different QoS policies have already been devised and implemented, maintaining
QoS in diverse scenarios adaptively and through continuous monitoring is still an
area of great concern and active research.

## References

1. International Telecommunication Union (ITU), The status of voice over internet protocol
   (VOIP) worldwide. New Initiatives Programme, Document: FoV/04, 12 Jan 2007 (2006)
2. Muhamad Amin AH, VOIP performance measurement using QoS parameters. In: The Second
   International Conference on Innovations in Information Technology, India (2005)
3. Quality of Service for Voice over IP, Cisco Systems (2001)
4. Diffserv—The scalable end-to-end quality of service model. Cisco Systems White Paper, Aug
   2005
5. Integrated Services. Available http://www.cisco.com
6. C. Aurrecoechea, A.T. Campbell, L. Hauw, A survey of QoS architectures. Multimedia Syst.
   6(3), 138–151 (1998)
7. Series G, Transmission systems and media, digital systems and networks. ITU-T
   Recommendation G.114, May 2003
8. H. Schulzrinne, S. Casner, R. Frederick, V. Jacobson, RTP: a transport protocol for real-time
   applications, IETF RFC 3550, July 2003
9. M. Bhatia, J. Davidson, S. Kalidindi, S. Mukherjee, J. Peters, VoIP: an in-depth analysis,
   Cisco Press (2006)
10. Methods for subjective determination of transmission quality, ITU-T Recommendation P.800,
    Aug 1996
11. Mean opinion score (MOS) terminology, ITU-T Recommendation P.800.1, Mar 2003
12. The E-model: a computational model for use in transmission planning, ITU-T
    Recommendation G.107, Dec 2011
13. Echo analysis for voice over IP, Cisco Systems, 2002
14. Y. Sung, C. Lund, M. Lin, S. Rao, S. Sen, Modeling and understanding end-to-end class of
    service policies in operational networks. In: Proceedings of SIGCOMM, Spain, (2009)
15. T. Koren, et al., Enhanced compressed RTP (CRTP) for links with high delay, packet loss and
    reordering, IETF RFC 3545, July 2003
16. C. Semeria, Supporting differentiated service classes: queue scheduling disciplines, Juniper
    Networks White Paper, Dec 2001
17. Comparing traffic policing and traffic shaping for bandwidth limiting, CISCO Systems (2005)
18. P. Hu, The impact of adaptive playout buffer algorithm on perceived speech quality
    transported over IP networks. M. Sc Thesis, University of Plymouth, Sept 2003
19. R. Ramjee, J. Kurose, D. Towsley, H. Schulzrinne, Adaptive playout mechanism for
    packetized audio application in wide-area networks, in *Proceedings of 13th IEEE Networking
    for Global Communications (INFOCOM)*, 1994

20. Technical Report on "Human factors (HF); quality of experience (QoE) requirements for real —time communication services" by ETSI, TR 102 643 V1.0.1, 2012
21. Report on "MDI/QoE for IPTV and VoIP; quality of experience for media over IP" by Ineoquest
22. B. Branden, et al., Recommendations on queue management and congestion avoidance in the internet, RFC 2309, April 1998
23. N. Leng, S. Hoh, D. Singh, Effectiveness of adaptive codec switching VoIP application over heterogeneous networks, in *2nd International Conference on Mobile Technology, Applications and Systems*, p. 7, China, 15–17 Nov 2005. https://doi.org/10.1109/mtas.2005. 207214
24. X. Lin, N.B. Shroff, R. Srikant, A tutorial on cross-layer optimization in wireless networks. IEEE J. Sel. Areas Commun. **24**(8), 1452–1463 (2006)
25. S. Khan, Y. Peng, E. Steinbach, M. Srgoi, W. Kellerer, Application-driven cross-layer optimization for video streaming over wireless networks. IEEE Commun. Mag. **44**(1), 122– 130 (2006). https://doi.org/10.1109/MCOM.2006.1580942
26. A. Panousopoulou, G. Nikolakopoulos, A. Tzes, J. Lygeros, Recent trends on QoS for wireless networked controlled systems. The Mediterranean J. Comput. Netw. **2**(1), 31–40 (2006)

# Chapter 4
# VoIP Over Wireless LANs—Prospects and Challenges

## 4.1 Introduction

Wireless LANs (WLANs) are increasingly becoming pervasive among enterprises. As wireless voice clients increase day by day, use of modern dual-mode (wireless and cellular) smartphones coupled with increased applications of wireless services has transformed WLAN from a mere source of convenience to an essential element for the network infrastructure. Therefore, considering WLANs as the primary network and deploying VoIP-based services over it require a thorough analysis of the design challenges and also demand critical problem-solving approaches to address the issues of QoS maintenance both for unlicensed wireless links as well as an unreliable IP platform.

## 4.2 Overview of Wireless LANs

WLAN standards can be grouped into several families. The IEEE 802.11 family is comprised of:

- 802.11a—Up to 54 Mb/s in the 5 GHz band, using OFDM3 modulation scheme and WEP4 and WPA5 security [1].
- 802.11b—Up to 11 Mb/s in the 2.4 GHz band, using DSSS-CCK6 modulation, and WEP and WPA security [2].
- 802.11g—Up to 54 Mb/s in the 2.4 GHz band, using OFDM or DSS with CCK modulation, and WEP and WPA security [3].

At the moment, 802.11b is probably the most widely used WLAN standard [4], but there are devices that are compatible with all three standards in the same time.

© Springer International Publishing AG, part of Springer Nature 2019
T. Chakraborty et al., *VoIP Technology: Applications and Challenges*,
Springer Series in Wireless Technology, https://doi.org/10.1007/978-3-319-95594-0_4

IEEE 802.11 defines two different architectures, basic service set (BSS) as shown in Fig. 4.1 and independent basic service set (IBSS). In a basic service set, wireless stations, called STAs, are associated with an AP.

All communications take place through the AP. In an Independent basic service set, STAs can communicate directly to each other, providing that they are within each other's transmission range. This form of architecture is facilitated to form a wireless ad hoc network in absence of any network infrastructure. Several BSS can be connected together via a distribution system (DS) to form an extended network, called extended service set (ESS).

IEEE 802.11 defines MAC and physical (PHY) layer specifications for WLANs. Three different physical layer specifications have been defined, namely frequency hopping spread spectrum (FHSS), direct sequence spread spectrum (DSSS), and infrared (IR), with the maximum data transmission rate of up to 2 Mbps. The DSSS and FHSS physical layers are operated in the license-free 2.4 GHz ISM band.

IEEE 802.11 MAC defines two different access mechanisms, the mandatory distributed coordination function (DCF) which provides distributed channel access based on carrier sense multiple access with Collision Avoidance (CSMA/CA) and the optional PCF which provides centrally controlled channel access through polling. In DCF, all stations contend for the access to the medium, in distributive manner, based on the CSMA/CA protocol. For this reason, the access mechanism is also referred to as contention-based channel access. In PCF, a point coordinator

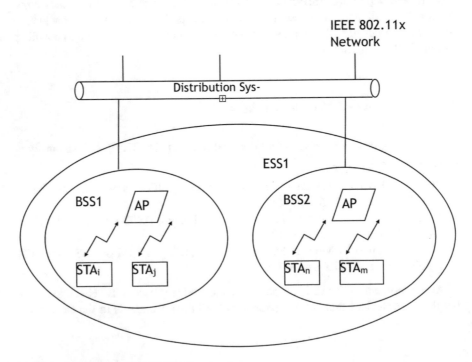

**Fig. 4.1** Architecture of IEEE 802.11 network

(PC) which is most often collocated in AP controls the medium access based on the polling scheme, such that the PC polls individual stations to grant access to the medium based on their requirements. As in PCF, stations do not contend for the medium. Instead, the medium access is controlled centrally and the access mechanism is sometimes referred to as contention-free channel access.

An additional mechanism, RTS/CTS, is defined to solve the hidden terminal problem found in wireless networks that use CSMA. With RTS/CTS, the sender and receiver perform a handshake mechanism by exchanging RTS and CTS control frames.

IEEE 802.11 has gained immense popularity due to its cost-effectiveness and easy deployment. Today, IEEE 802.11 hotspots are available at offices, campuses, airports, hotels, public transport stations, and residential places, making it one of the most widely deployed wireless network technologies in the world.

## 4.3 Important Aspects Regarding Voice Over WLAN

Given the importance of deploying VoIP both as service and an application in Wireless LANs, there are few important issues that one must be aware of before deploying VoIP or replacing existing telephony services with VoIP. Currently, some wireless phone companies (carriers) are gearing up to offer hybrid services, where the phone will use VoIP over Wi-Fi when it is available, and subsequently switch to cellular connection, when the Wi-Fi is not available. So, depending on the context, VoIP over WLAN encompasses different aspects, some of which are discussed here.

1. Wireless VoIP is more advantageous over cellular networks

A big advantage of VoIP is that modern cell phones which are mostly IP enabled and can easily connect to Wi-Fi networks often provide a free alternative solution to communication services. This is because, nowadays, there are numerous free 802.11 hotspots free for use by the common public and VoIP can certainly take advantage of this issue. Moreover, using VoIP can also reduce the need to own different SIM cards while roaming to different countries, significantly saving the costs and the troubles associated with it.

2. VoIP over WLAN has several uses

VoIP over wireless LAN can offer a cost-effective and easy alternative solution to connect different campuses together in public and private enterprises. It can provide easy internal calling for corporations, educational campuses, hospitals, hotels, government buildings, and multiple-tenant units such as dorms, with the ability to roam freely and advanced calling features such as voicemail and caller ID. As an added benefit, users can use the LAN's Internet connection and an account with a VoIP provider to make calls outside the site, including domestic long-distance and international calls, often at no extra charge.

3. WiMAX can enhance the VoIP services over WLAN

WiMAX is a developed technology based on the IEEE 802.16 standards and follows the long-range microwave-based wireless system to provide extensive broadband coverage to a large rural area or an entire metropolitan city. Considering its span of around 76 km with theoretical and practical throughputs around 280 and 70 Mbps, respectively, WiMAX has the bandwidth to support VoIP and can extend the reach of VoIP services to more people.

4. VoIP is more sensitive than data transmission

Considering the real-time QoS constraints of VoIP as elaborately described in Chap. 3, the wireless network where VoIP will be deployed must be reliable enough to support the QoS guarantees with respect to the end-users. Therefore, mixing of VoIP and competing data packets in a common network can often degrade the VoIP call quality, unless appropriate measures are taken to curb this issue.

5. Security is a major concern for wireless networks

The vulnerabilities in the wireless platform can pose a serious question with respect to the security aspect in VoIP communication. On top of that, IP platform is a best-effort traffic network and has its own set of security threats. These become more prominent when the call signaling protocols are deployed for VoIP, like SIP and H.323 (as already discussed in the previous chapters). Unless privacy of communication is upheld, VoIP will fall in confidence for the existing subscribers.

6. VoIP supports limited backward compatibility

There is no denying the fact any IP enabled device will support VoIP. However, older equipment which is not meant to handle the real-time requirements of VoIP will falter in providing QoS for the same. This is because considering proprietary systems, it is not possible to modify their hardware/software specifications for making them compatible with VoIP requirements. At the same time, legacy users will also face isolation as VoIP vendors continue to upgrade their software for newer equipment with advanced OS features.

7. Smartphones combine cellular and Wi-Fi VoIP

Modern smartphones and related communication devices are introspecting the idea of seamless switching between cellular and Wi-Fi VoIP within the same call, based on the availability of WLAN. This will also drive the landlines to an obsolete category, as these hybrid phones will provide the users with an all-encompassing solution: Use Wi-Fi network at home and use cellular technology when there is no Wi-Fi within range.

Currently, some wireless phone companies (carriers) are gearing up to offer hybrid services, where the phone will use VoIP over Wi-Fi when it is available, and subsequently switch to cellular connection, when the Wi-Fi is not available.

## 4.4 Design Challenges for VoIP Services in WLAN

The first and foremost step before successfully implementing VoIP in WLAN is to address the various challenging aspects that the unification of these two technologies presented and highlighted briefly.

### 4.4.1 VoIP Over Evolving WLAN Standards

While the WLAN was initially developed keeping in mind the data communications, significant enhancements have been designed and are also being proposed with respect to the various standards of WLAN. In other words, WLAN is evolving with time. Now this can prove to be a serious hindrance to the VoIP over WLAN, not only because VoIP has to cope with this evolution through modifications in its signaling and call management policies, but also due its stringent QoS limitations (which are minimal for normal data transmissions). Also, it must be kept in mind that even though WLAN standards and VoIP continue to work in different layers and apparently seem to be independent of each other, actually they are tightly coupled in the sense that any loss of signal information can subsequently increase the packet loss for VoIP calls, which is highly undesirable [5].

### 4.4.2 System Capacity in WLAN

The total system capacity in WLAN has a bearing on the achieved QoS for VoIP calls. While system capacity is defined as the total bandwidth available for transmissions in a network (measured in rate per second), with respect to VoIP, the system capacity can be defined as the "total number of simultaneous VoIP calls allowed in the basic service set (BSS)."

Obviously, application vendors will opt for higher system capacity to increase the reach of their VoIP services to the customers. However, this system capacity is also related with the QoS for VoIP calls. This is inherent in the definition of QoS (i.e., to provide agreed Quality of Service in a network when the "system capacity"

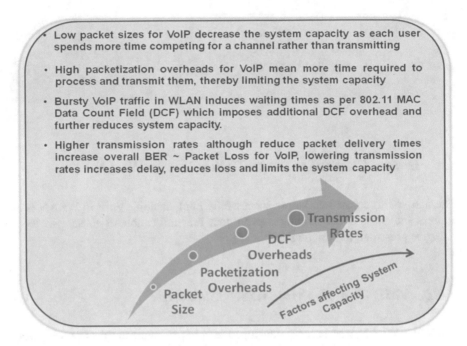

**Fig. 4.2** Relation between WLAN system capacity and VoIP QoS

is low). There is no need to worry for QoS if the available system capacity is much more than the VoIP thresholds. Figure 4.2 illustrates how the system capacity varies with respect to different aspects of VoIP technology (including packet size, packetization and DCF overheads, transmission rates).

The total system capacity in WLAN has a bearing on the achieved QoS for VoIP calls.

### 4.4.3   Inadequacy of PCF

It is already found out that DCF has major drawbacks (especially keeping VoIP in mind). WLAN has been modified to overcome this issue with the incorporation of point coordination function (PCF) which divides time into superframes. Each superframe is further subdivided into a contention-free period (CFP) followed by a contention period (CP). Thus, a CFP and a CP occur alternately with the constraint that the CP must be long enough to deliver at least one MSDU. However, PCF takes into account the number of MSDUs to be sent (and not the total time required to transmit the MSDUs). Therefore, this aspect will have critical implications for VoIP when they are competing with data traffic whose payload size is always higher.

### 4.4.4   Admission Control Issues

Generally, admission control in any network can be based on two policies, namely (i) authentication-based (where the user is allowed in the network only after authenticating its proof of being a valid user) and (ii) resource-based (where a user is inducted in the network only when the system knows it can allocate requisite resources for that user). The second option suits QoS and must be followed for VoIP over WLAN. However, two factors proved to be a hindrance in this regard [6].

 (i)   WLAN designed mainly for data communication does not inherently provide resource-based admission control, unless otherwise enhanced.
(ii)   Since the total system capacity is severely limited by VoIP calls as already discussed, reserving resources for VoIP even using external means is not a feasible solution in such scenarios.

### 4.4.5   Security Challenges

It is already known that the unreliable IP links are subject to various security issues including snooping, denial-of-service attacks that can effectively hamper the privacy aspects of communication and leave all the customers dissatisfied. Although security has been a source of major research in WLANs, their suitability with respect to real-time traffic must be properly evaluated before implementation.

### 4.4.6   Sleep Mode for WLAN and Power-Saving Issues

WLAN applications normally send a beacon period after periodic intervals and use the sleep modes to save power. However, this means that based on the beacon transmission, all the pending packets will be stored in the buffer (e.g., of the access points). One way to reduce this waiting delay is to shorten the beacon interval time. This, in turn, indicates that when not in use, a Wi-Fi enabled handset must wake up more often for beacons, significantly reducing the available power in the process.

### 4.4.7  Handoff Problem in WLAN

As VoIP users move from one AP to another AP in a WLAN, handoff occurs, which is also called "roaming" [6] in WLAN terminology. Ideally, as VoIP operates in the upper layers, this handoff should be transparent to the end-users who must not face any significant interruption due to handoff. This issue of transparency during handoff is not a problem for data applications, as the underlying TCP protocol can conceal all the lost packets, and using retransmission provides a reliable service. On the other hand, VoIP mostly uses UDP to reduce the delay. UDP being unreliable cannot conceal this packet loss and also the associated handoff delay during roaming. Hence, mobility issues must also be thoroughly taken care of for VoIP communication in WLAN.

> It must be kept in mind that even though WLAN standards and VoIP continue to work in different layers and apparently seem to be independent of each other, actually they are tightly coupled in the sense that any loss of signal information can subsequently increase the packet loss for VoIP calls, which is highly undesirable.

## 4.5  Prospective Solutions for VoIP Policy Design in WLAN

Considering the discussed issues and challenges for VoIP applications in WLAN, probable solutions must aim to include some critical functionality in the developed algorithms to be implemented in the compatible software and hardware [7, 8]. These essential solution aspects are discussed henceforth.

1. *QoS maintenance following a QoS priority mechanism* especially in heterogeneous networks, where this QoS priority must be enforced upon using a QoS-based call admission control policy.
2. *The ability to differentiate VoIP traffic from nonreal-time IP traffic,* i.e., classifying traffic based on their source information and subsequently optimizing this VoIP traffic stream so as to increase reliability even in the presence of unreliable transport layer protocol (UDP).
3. *Incorporating security features including both authentication and encryption* to ensure data privacy and integrity, without obviously compromising with the voice quality.
4. *Seamless cross-layer-based approach* to handle the complex issues of mobility including handoff, so as to reduce both handoff delay and packet loss during handoff.

5. *Design of an effective monitoring mechanism* that can proactively identify the VoIP performance issues and rectify them during early stages.
6. *WLAN management including parameterization and policy formulation* for the network elements (access points, routers, relay nodes, etc.) so as to increase both the network coverage and overall system capacity.
7. *Power management features for VoIP phones* by exploiting the features in both domains with respect to WLAN (sleep mode, PCF, etc.) and VoIP (silent suppression codecs, voice activity detector, etc.).

## 4.6   Related Works

Maintaining the QoS for VoIP applications has witnessed extensive research activity following the rapid deployment of VoIP technology in various networks. Cross-layer optimization techniques [9], for example, have been proposed for improving the VoIP performance particularly under congested scenarios. Currently, adaptive QoS optimization techniques [10, 11] are also designed to deal with various unpredictable disturbances in WLANs. Apart from optimizations in the links and traffic, the APs need to be properly optimized in respect of queue length [12], retry limit [13], RTS threshold [14], etc.

Selection of APs has principally been focused when a node is in the vicinity of multiple APs. Selection based on the packet transmission delay associated with each AP has been considered in [15]. De-association of low throughput node is also a convenient technique to dynamically select AP in congested WLAN [16]. APs sending beacons for transmitting the R-values to all nodes in the wireless network enable the selection of best AP [17]. Selection based on a decision metric involving real-time throughput has been considered in [18] both for static and dynamic selection algorithm. AP association protocols based on throughput and frame loss rate [19, 20] have been considered as selection criteria.

For significant improvement in VoIP performance, apart from the selection of APs, stress has also to be given in controlling various parameters associated with the APs including the antennae. Significant optimization has been done in the domain of AP antennae for enhanced performance of VoIP networks [21, 22]. A scheme for increasing the number of simultaneous voice calls giving proper priority to APs has been highlighted in [23] with degradation of QoS well within the practically acceptable limit. The concepts of live migration and virtual AP [24] have also been suggested for the efficient deployment of APs in WLANs. Modified Deficit Round Robin (DRR) algorithm [25] replacing the standard point coordination function (PCF) polling algorithm is the premium consideration behind the improvement in VoIP performance.

However, suitable analysis of the packet access delay, based on M/G/I model, provides an improved PCF algorithm [26] for better VoIP performance. Efficient handover to different APs [12] solves the problem of bottleneck in a highly congested network. A limit to maximum number of admission calls could be derived using transmission opportunity (TXOP) parameter of the media access control (MAC) protocol [27].

Advanced search techniques as proposed in [28 29] are currently being applied to handle the real-time traffic. Most of the previous state-space search approaches have been directed toward finding the optimal routes between intermediate nodes in various network conditions as in [30]. However, little work has been done with respect to mapping a particular network-related problem, other than finding shortest routes, to state-space problem and solving it. For instance, cell-to-switch assignment is developed using heuristic search for mobile calls in [31]. Heuristic search is further used in [32] to simulate possible attackers searching for attacks in modeled network for proactive and continuous identification of network attacks.

Adapting parameters of VoIP flows (like codec or packetization interval used) have also been an area of active research and have witnessed considerable development. Evaluation of congestion due to rate change is analyzed in [33], and codec change is proposed for the node suffering rate change. In [34], when the channel is detected congested (based on real-time control protocol (RTCP) packet loss and delay feedback information), a central element performs codec adaptation using common transcoding methods for calls entering wireless cell. Focus shifts from multirate effects to congestion provoked by existence of additional VoIP sessions on the cell in [35] where the algorithm focuses on adapting low-priority calls in favor of high-priority ones by "down-switching" codec and packetization interval of low-priority calls based on a set of information.

Thus, the literature survey points to the active interest among the research community toward delivering feasible telephony solutions involving VoIP technology across different networks.

> Maintaining the QoS for VoIP applications has witnessed extensive research activity following the rapid deployment of VoIP technology in various networks, including the formulation of cross-layer and adaptive QoS optimization techniques.

## 4.7  Test-Bed Model for VoIP Deployment Over WLAN

In order to study the deployment issues including the challenges in the physical, data link, and application layers for VoIP applications in WLAN, a test-bed model is developed to conduct several performance analysis tests for successful VoIP communication over WLAN. This model will also serve as the platform for implementing various optimizations in the design and development processes, as discussed in the subsequent chapters.

The VoIP test bed is shown in Fig. 4.3. It consists of a software server, a hardware server, a mobility controller, few computers, Nokia E51 mobile phones, switches, APs, and necessary software for managing and monitoring VoIP calls.

There are two switches controlling the APs and other equipment. These two switches are in different networks and connected by bridging. Such a connection setup results in the creation of two scenarios, namely wired local area network (LAN) and WLAN.

In the wired LAN scenario, all packets are routed by switch 1 to the appropriate destination. In the WLAN setup, switches 1 and 2 along with the APs participate in routing the packets. The mobility controller provides mobility by virtue of which mobile nodes can communicate with each other while roaming. The Brekeke Software server is SIP proxy and registrar server and manages all VoIP sessions in both the scenarios.

## 4.7.1 Hardware Modules

### A. Brekeke SIP Server

The Brekeke SIP server is an open standard-based SIP proxy server and registrar. It authenticates and registers user agents such as VoIP device and soft phone and routes SIP sessions such as VoIP calls between user agents. When caller and callee are located on different networks, the Brekeke SIP server can connect calls by rewriting SIP packets appropriately, thereby supporting network address translation

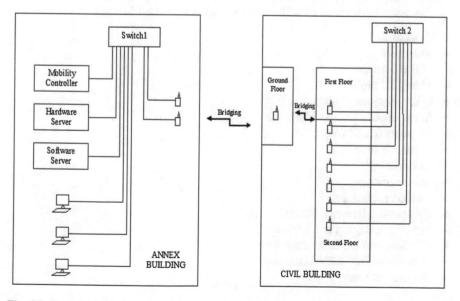

**Fig. 4.3** Domain diagram

(NAT) traversal. Upper/thru registration is another unique feature that allows easy configuration of parallel users of pre-existing or other SIP servers [36]. The current version deployed in the test bed is Ver. 2.2.7.8/276.

Brekeke SIP server has the following features:

- Registrar service
- Call routing
- NAT traversal
- Dial plan
- Upper/thru registration
- Authentication
- Session management
- Transmission control protocol (TCP) support
- Back-to-back user agent (B2BUA) mode
- IPv6
- DNS SRV
- TLS (advanced)
- RADIUS (advanced)

B.  MITEL Hardware Server

The Mitel 3300 server [37] provides the reliability and comprehensive features of a PBX. It's easy-to-use features and cost effectiveness for a small business solution, complete with productivity-enhancing applications and data networking capability make it an apt one for VoIP communications. Its functional modules are involved primarily in integrated voice mail, automatic call distribution, data networking, conferencing, paging, auto attendant, etc.

The essential features of Mitel 3300 server are enlisted henceforth.

- Remote working
- Voicemail
- Call recording
- Mitel UC mobile
- Mitel dynamic extension
- Conferencing and collaboration
- Phonebook and directory
- Music on hold
- Hot desking
- Multi-tenancy
- Unified messaging
- Computer telephony integration (CTI)
- Instant messaging and presence
- Wireless and premises mobility
- Formal contact centers/call centers
- Informal contact centers

## C.  ProCurve Mobility Controller

It is used to provide centralized management, control, and configuration of the APs. Robust identity and role-based user account profiles, as well as virtual service communities (VSCs) with independently configurable QoS, authentication, encryption, and virtual local area network (VLAN) support, deliver intelligence to the edge of the network [38].

Some of the benefits of using this controller are listed below.

- Ease of use, scalability, and redundancy
- Enhanced architecture for flexible network design
- IEEE 802.11a/b/g/n AP and access device support
- Comprehensive WLAN security
- Appliance and blade form factors

## D.  ProCurve Switch

The ProCurve 3500yl-24G-PWR [39] is an advanced layer 3 switch. It has 20 10/100/1000Base-T ports, four dual-personality ports, integrated PoE on all 10/100/1000Base-T interfaces, and an expansion slot for an optional 4-port 10-GbE module. The foundation for this switch is a purpose-built, programmable ProVision ASIC that allows the most demanding networking features such as QoS and security to be implemented in a scalable yet granular fashion. It offers flexibility and scalability, as well as ease of deployment, operation, and maintenance for any network environment. The basic features of this switch are described as follows.

- Stacking Capability
- Resiliency and high availability (using virtual router redundancy protocol, 802.1s multiple spanning tree protocol, etc.)
- ProCurve switch meshing, VLAN support, support for group VLAN registration protocol
- Static IP routing, support for RIP, OSPF protocols.
- Security (switch CPU protection, virus throttling, etc.)
- Multiple user authentication, authentication flexibility
- QoS support (traffic prioritization, bandwidth shaping, class of service)

## E.  ProCurve Access Point

HP ProCurve MSM310 [40] APs play in important role in managing the wireless calls. It has a single radio with two antennas for diversity. The antenna connectors are reverse-polarity SMA jacks. Antennas can be mounted directly on the MSM310, or an external antenna can be connected to the main antenna connector. The MSM310 can operate in one of two modes, namely controlled (the default) or autonomous. In the controlled mode, the MSM310 must establish a management tunnel with an MSM7xx Controller via Ethernet Port 1 to become operational. The controller manages the MSM310 and provides all configuration settings. Once switched to autonomous mode, MSM310 operates as stand-alone AP that is configured and managed via web-based management tool. The key benefits of this model are listed as follows.

- Single, dual and tri-radio
- Intelligent traffic forwarding, built-in security
- Self-healing, advanced mesh capabilities
- Client access
- Indoor and outdoor enclosures

F.  Nokia E51 Mobile Phone

Nokia E51 [41] mobile phones are used in the test bed to enable wireless VoIP calls. This phone supports GPRS, EDGE, 3G, WLAN, and Bluetooth. It has Symbian OS (9.2 version) with ARM 11 369 MHz processor. It can operate in the 2G network (GSM 850/900, 1800, 1900) and 3G network (HSDPA 850/2100) and supports SIP as the call signaling protocol in VoIP communications.

### 4.7.2   Software Elements

A.  X-Lite

X–Lite 3.0 is CounterPath's next-generation soft phone client, offering users all the productivity of a traditional telephone with desktop and mobile computer enhancements. It has all the standard telephone features, including call display and message waiting indicator (MWI), speakerphone, mute, redial, hold, do not disturb, call ignore, call history (list of received, missed, dialed, and blocked calls), call forward, call record, three-way audio and video conferencing, acoustic echo cancelation, automatic gain control, voice activity detection and support for various audio and video codecs [42].

B.  ManageEngine VQManager

ManageEngine VQManager Ver. 6.1 is a powerful, web-based, $24 \times 7$ real-time QoS monitoring tool for VoIP networks. It enables IT administrators to monitor their VoIP network for voice quality, call traffic, bandwidth utilization, and keep track of active calls and failed calls. VQManager can monitor any device or user agent that supports SIP, Skinny, H.323 and RTP/RTCP [43].

C.  Network Emulator for Windows Toolkit (NEWT)

NEWT [44] is a software-based emulator that can emulate the behavior of both wired and wireless networks using a reliable physical link, such as Ethernet. Network attributes such as latency, amount of available bandwidth, queuing behavior, packet loss, reordering, and error propagations are incorporated in NEWT. It also provides flexibility in filtering packets based on IP address or protocols such as TCP, UDP, and Internet Control Message Protocol (ICMP).

When compared to actual hardware test beds, it is a cheaper and a more flexible solution in testing network-related software under various network conditions. The current version is 2.1.0003.0.

D. NetSim

NetSim [45] is used as the simulation tool for analysis purpose. The advantage of using NetSim is that it is simple to use and it can map the test-bed scenario. Therefore, the results of this simulation tool can be directly fed into the test bed. NetSim allows one to vary parameters of APs that include buffer size, retransmission limit, RTS–CTS (Clear To Send) threshold, and transmitter power. The factors related to environment can also be modified in this tool by varying parameters for path-loss, fading, shadowing, etc. The communicating nodes can further be simulated with the help of parameters that include buffer size, retransmission limit, RTS–CTS threshold, and codec properties.

> The test-bed model for deploying VoIP over WLAN comprises of several hardware and software modules and will serve as the platform for implementing various optimizations in the design and development processes.

## 4.8  Performance Analysis in the Test Bed

### 4.8.1  Analysis of Call Signaling Protocols

Initially, both SIP and H.323 (as described in Chap. 2) are implemented in wired scenarios created in the test bed as described in the previous section. The mean QoS metrics for the calls made by implementing SIP and H.323 are shown in Table 4.1. It is observed that the packet loss, MOS, and R-factor remain almost same in both the scenarios for real-time traffic while delay and jitter are comparable. This is because in the absence of any network congestion or other sources of link disruption, both SIP and H.323 perform equally well in managing VoIP sessions with acceptable call quality.

**Table 4.1** QoS metrics for SIP and H.323

| Metrics | SIP | H.323 |
|---|---|---|
| Delay (ms) | 10 | 4 |
| Jitter (ms) | 14 | 7 |
| Loss (%) | 0 | 0 |
| MOS | 4.4 | 4.4 |
| R-factor | 93 | 93 |

In absence of any network congestion or other sources of link disruption, both SIP and H.323 perform equally well in managing VoIP sessions with acceptable call quality.

The call signaling protocols are thereafter analyzed by implementing VoIP calls in the wireless domain. The QoS data is obtained by taking the readings of the parameter values over two wireless handsets with a call running from one handset to the other. Few calls are made, and the obtained values are considered for analysis. As observed from Fig. 4.4, the average packet loss is more in SIP than in H.323. In the absence of any network congestion, such loss can be attributed to environmental factors like path-loss, fading, and shadowing that play a major role in IEEE 802.11b LANs. Figures 4.5 and 4.6 show the variation of MOS and R-factor for various access points. It is observed that with increase in distance, the call quality is high with respect to MOS and R-factor for SIP than H.323. The graphs also reveal that though H.323 has a large initial degradation in both MOS and R-factor value, it provides a constant performance though at low levels independent of the number of access points. On the other hand, the MOS and R-factor value of SIP have degraded with increase in distance.

With increase in distance, the call quality is high with respect to MOS and R-factor for SIP than H.323.
   Though H.323 has a large initial degradation in both MOS and R-factor value, it assumes a constant performance though at low levels independent of the number of access points, unlike SIP which experiences degradation.

Thus, selection of appropriate call signaling protocols plays an important role toward QoS maintenance of ongoing VoIP calls in wired and wireless domains. The choice is dependent on the server and the soft phones used, the prevailing network conditions, and related environmental factors.

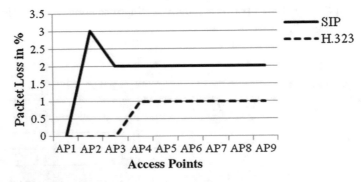

**Fig. 4.4** Variation in packet loss for every access point

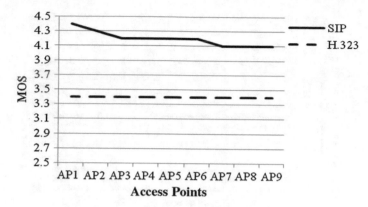

**Fig. 4.5**  Variation in MOS for every access point

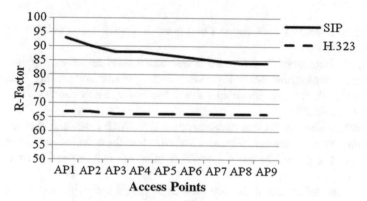

**Fig. 4.6**  Variation in *R*-factor for every access point

## 4.8.2   *Analysis of Codec Parameters and Buffer Size*

Each VoIP entity that includes server, soft phone, and access point has a buffer which stores and forwards packets based on a set of certain rules. The optimal buffer size depends on the packets per second generated and hence is dependent on the codec bit rate. Therefore, selection of proper buffer size in conjunction with the codec bit rate is of crucial importance toward maintenance of appropriate call quality.

Initially, the calls are implemented in wired scenarios as created in the test bed (described in Sect. 4.7). Network congestion is introduced in the link in the form of background traffic with the help of NEWT emulator. Initially, a high bit rate codec is chosen for existing calls, and the background traffic is increased slowly. As observed from Fig. 4.7, both delay and packet loss have increased and crossed the

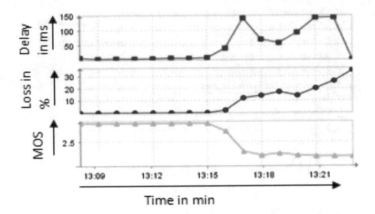

**Fig. 4.7** Variation in delay, loss, and MOS with time for high bit rate codec

threshold limit. MOS values suggest that the call quality degrades with increase in network congestion.

Then, we implement low bit rate codec-based calls under equivalent network conditions. As observed from Fig. 4.8, all the parameters namely latency and packet loss are below the threshold values. Moreover, MOS degrades but remains within tolerable limits.

Thereafter, adaptive bit rate codec is applied to VoIP calls, and similar network congestion levels are created. Figure 4.9 shows that there are negligible delay and packet loss. The overall MOS is 4.4 which indicates that the call is of very high quality.

Finally, the buffer size of the soft phones is varied and calls are made for high, low, and adaptive bit rate codecs. As seen from Fig. 4.10, packet loss increases sharply as buffer size is decreased from 5 to 1 MB in case of high bit rate codec.

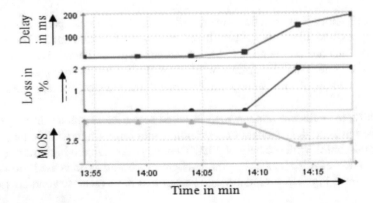

**Fig. 4.8** Variation in delay, loss, and MOS with time for low bit rate codec

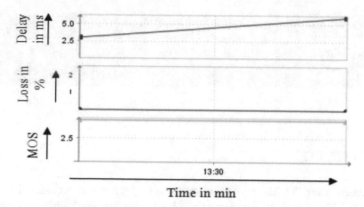

Fig. 4.9 Variation in delay, loss, and MOS with time for adaptive bit rate codec

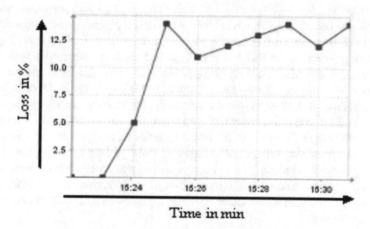

Fig. 4.10 Variation in packet loss with decrease in buffer size over time

However, low and adaptive bit rate codecs record negligible packet loss under identical network scenarios.

Therefore, codec bit rate must be selected properly based on the network conditions and the call quality as promised by the operator. The associated buffer size is dependent on the selected codec bit rate and further affects the overall performance of VoIP calls. Adaptive bit rate ensures maximum QoS enhancement and requires minimum queue size as compared to high bit rate codecs. Low bit rate codecs must be implemented only when the network is severely constrained, and no adaptive bit rate codec is available. However, the final selection always depends on the acceptable call quality as per the demands of the end-user.

Selection of proper buffer size in conjunction with the codec bit rate is of crucial importance toward maintenance of appropriate call quality IN VoIP applications.

## 4.9 Summary

VoIP studies over WLAN are carried out in this chapter that includes the different aspects of WLAN technology, design challenges toward deploying real-time VoIP services, and finally, some general approaches related to VoIP deployment over WLAN. This is followed by a comprehensive real-time study of VoIP communication over WLAN in a generic test bed. The various software and hardware elements are described along with the test-bed setup that spans across multiple buildings. Extensive performance analysis of important factors affecting VoIP services is carried out in this test bed. Test-bed readings verify the fact that the selection of appropriate call signaling protocols, codec bit rate, and associated buffer size is an important step toward attaining high QoS for VoIP calls. Diverse network scenarios further trigger the necessity for adaptive modifications during ongoing VoIP sessions as clearly reflected from the test-bed outcome. This test bed will further serve as the platform for various studies on analysis and implementation of different optimizing techniques to improve VoIP performance in general. To be specific, parameter configuration of servers and related VoIP nodes and subsequent test-bed implementation of QoS optimization algorithms can only be realized after addressing the relevant issues affecting VoIP applications and the same is discussed in the next chapter.

## References

1. IEEE Std. 802.11a, *Supplement to Part 11: Wireless LAN Medium Access Control (MAC) and Physical Layer (PHY) Specifications: Higher-Speed Physical Layer Extension in the 5 GHz Band* (1999)
2. IEEE 802.11b, *Wireless LAN Medium Access Control (MAC) and Physical (PHY) Layer Specification: High Speed Physical Layer Extensions in the 2.4 GHz Band, Supplement to IEEE 802.11 Standard* (September 1999)
3. IEEE Std. 802.11g, *Supplement to Part 11: Wireless LAN Medium Access Control (MAC) and Physical Layer (PHY) Specifications: Further Higher-Speed Physical Layer Extension in the 2.4 GHz Band* (2003)
4. R. Beuran, *VoIP Over Wireless LAN Survey*, Research Report, IS-RR-2006-005, Japan Advanced Institute of Science and Technology (JAIST), Ishikawa, Japan (April 2006)

5. L. Cai, Y. Xiao, X. Shen, L. Cai, J.W. Mark, Control mechanisms for packet audio in the Internet. Int. J. Commun. Syst. 498–508 (2006). https://doi.org/10.1002/dac.801. (Wiley InterScience)
6. P. Chandra, D. Lide, *Wi-Fi Telephony Challenges and Solutions for Voice Over WLANs* (Elsevier, 2006)
7. W. Wei, L.C. Soung, V.O.K. Li, Solutions to performance problems in VoIP over a 802.11 wireless LAN, in IEEE Trans. Veh. Technol. **54**(1), 366–384 (2005). https://doi.org/10.1109/tvt.2004.838890
8. White Paper on *"Is Your WLAN Ready for Voice?"* (CISCO Press, 2008)
9. Q. Zhang, Y. Zhang, Cross-layer design for QoS support in multihop wireless networks. Proc. IEEE **96**(1), 64–76 (2008). https://doi.org/10.1109/JPROC.2007.909930
10. Z. Qiao, L. Sun, N. Heilemann, E. Ifeachor, A new method for VoIP Quality of service control using combined adaptive sender rate and priority marking, in *2004 IEEE International Conference on Communications*, vol. 3, pp. 1473-1477, France, June 2004. https://doi.org/10.1109/icc.2004.1312756
11. S. Shin, H. Schulzrinne, Balancing uplink and downlink delay of VoIP traffic in WLANs using adaptive priority control (APC), in *Proceedings of the 3rd International Conference on Quality of Service in Heterogeneous Wired/Wireless Networks, QShine '06*, Canada. https://doi.org/10.1145/1185373.1185426
12. M. Niswar et al., Seamless VoWLAN handoff management based on estimation of AP queue length and frame retries, in *IEEE International Conference on Pervasive Computing and Communications, PerCom 2009*, pp. 1–6, 9–13, USA, Mar 2009. https://doi.org/10.1109/percom.2009.4912877
13. N. Kim, H. Yoon, Packet fair queueing algorithms for wireless networks with link level retransmission, in *Proceedings of the IEEE Consumer Communications and Networking Conference (CCNC '04)*, pp. 122–127, USA, Jan 2004. https://doi.org/10.1109/ccnc.2004.1286844
14. S.M.R. Hasan, M.S. Islam, N. Hasan, A. Rahman, Exploiting packet distribution for tuning RTS threshold in IEEE 802.11, in *25th Biennial Symposium on Communications (QBSC)*, pp. 369–372, Canada, 12–14 May 2010. https://doi.org/10.1109/bsc.2010.5472955
15. V. Siris, D. Evaggelatou, Access point selection for improving throughput fairness in wireless LANs, in *10th IFIP/IEEE International Symposium on Integrated Network Management, IM '07*, pp. 469–477, Germany, May 21 2007–Yearly 25 2007. https://doi.org/10.1109/inm.2007.374812
16. E. Ghazisaeedi, S. Zokaei, A method for access point selection in 802.11 networks, in *First International Conference on Networked Digital Technologies, NDT '09*, pp. 447–451, The Czech Republic, 28–31 July 2009. https://doi.org/10.1109/ndt.2009.5272809
17. M. Tuysuz, H. Mantar, Access point selection and reducing the handoff latency to support VoIP traffic, in *2010 International Conference on Computer Engineering and Systems (ICCES)*, pp. 58–63, Egypt, Nov 30 2010–Dec 2 2010. https://doi.org/10.1109/icces.2010.5674897
18. M. Abusubaih, J. Gross, S. Wiethoelter, A. Wolisz, On access point selection in IEEE 802.11 wireless local area networks, in *Proceedings 2006 of 31st IEEE Conference on Local Computer Networks*, pp. 879–886, USA, 14–16 Nov 2006. https://doi.org/10.1109/lcn.2006.322194
19. F. Xu, C. Tan, Q. Li, G. Yan, J. Wu, Designing a practical access point association protocol, in *Proceedings of IEEE INFOCOM*, USA, pp. 1–9, 14–19 March 2010. https://doi.org/10.1109/infcom.2010.5461909
20. Y. Zhu et al., A multi-AP architecture for high-density WLANs: protocol design and experimental evaluation, in *5th Annual IEEE Communications Society Conference on Sensor, Mesh and Ad Hoc Communications and Networks, SECON '08*, pp. 28–36, USA, 16–20 June 2008. https://doi.org/10.1109/sahcn.2008.14

21. Q. Tan, Z. Zhang, M. Qiu, Y. Xu, A novel WLAN access point antenna with omnidirectional coverage, in *IEEE International Symposium on Antennas and Propagation Society*, vol. 3A, pp. 491–494, USA, 3–8 July 2005. https://doi.org/10.1109/aps.2005.1552294

22. F. Cladwell, Jr, J. Kenney, M. Ingram, Design and implementation of a switched-beam smart antenna for an 802.1 lb wireless access point, in *IEEE Radio and Wireless Conference, RAWCON 2002*, USA, pp. 55–58, 2002. https://doi.org/10.1109/rawcon.2002.1030116

23. D. Hashmi, P. Kiran, B. Lall, Access point priority based capacity enhancement scheme for VoIP over WLAN, in *Annual IEEE India Conference*, pp. 1–4, India, 15–17 Sept 2006. https://doi.org/10.1109/indcon.2006.302853

24. T. Hamaguchi, T, Komata, T, Nagai, H. Shigeno, A framework of better deployment for WLAN access point using virtualization technique, in *IEEE 24th International Conference on Advanced Information Networking and Applications Workshops (WAINA)*, pp. 968–973, Australia, 20–23 April 2010. https://doi.org/10.1109/waina.2010.61

25. W. Quan, D. Hui, Improving the performance of WLAN to support VoIP application, in *2nd International Conference on Mobile Technology, Applications and Systems*, p. 5, China, 15–17 Nov 2005. https://doi.org/10.1109/mtas.2005.207204

26. W. Quan, M. Du, Queueing analysis and delay mitigation in the access point of VoWLAN, in *IEEE International Symposium on Communications and Information Technology, ISCIT 2005*, vol. 2, pp. 1160–1163, China, 12–14 Oct 2005. https://doi.org/10.1109/iscit.2005.1567075

27. K. Stoeckigt, H. Vu, VoIP capacity—analysis, improvements, and limits in IEEE 802.11 wireless LAN. IEEE Trans. Veh. Technol. **59**(9), 4553–4563 (2010). https://doi.org/10.1109/TVT.2010.2068318

28. S. Koenig, M. Likhachev, Y. Liu, D. Furcy, *Incremental heuristic search in artificial intelligence*, AI Magazine, vol. 25 Issue 2, pp. 99–112, Summer 2004

29. S. Koenig, Real-time heuristic search: research issues, in *International Conference on Artificial Intelligence Planning Systems*, Pittsburgh, Pennsylvania, June 1998

30. T. Zhu, W. Xiang, Towards optimized routing approach for dynamic shortest path selection in traffic networks, in *International Conference on Advanced Computer Theory and Engineering*, pp. 543–547, Thailand, 20–22 December 2008

31. S. Mandal, D. Saha, A. Mahanti, Heuristic search techniques for cell to switch assignment in location area planning for cellular networks, in *IEEE International Conference on Communications*, vol. 7, pp. 4307–4311, France, 20–24 June 2004

32. V. Franqueira, Finding Multi-Step Attacks in Computer Networks Using Heuristic Search and Mobile Ambients, Ph.D. Dissertation, University of Twente, Netherlands, 2009

33. P. McGovern, S. Murphy, L. Murphy, Addressing the link adaptation problem for VoWLAN using codec adaptation, in *Proceedings of the Global Telecommunications Conference (GLOBECOM '06)*, USA, Dec 2006

34. A. Trad, Q. Ni, H. Afifi, *Adaptive VoIP transmission over heterogeneous wired/wireless networks*. Interactive Multimedia and Next-generation Networks. Lecture Notes in Computer Science, vol. 3311 (2004), pp. 25–36

35. B. Tebbani, K. Haddadou, Codec-based adaptive QoS control for VoWLAN with differentiated services, in *Proceedings of the 1st IFIPWireless Days (WD '08)*, France, November 2008

36. Brekeke Wiki, Available at http://wiki.brekeke.com

37. The Mitel 3300 Documentation, Available at http://www.mitel.com/DocController?documentId=10141

38. ProCurve MultiService Controller Series Overview, Available at http://procurve.com

39. The ProCurve 3500yl-24G-PWR Manual, Available at www.hp.com/rnd/pdfs/.../ProCurve_Switch_3500yl-24G-PWR.pdf

40. HP ProCurve MSM310 Manual, Available at http://www.hp.com/rnd/support/manuals/mapoutseries.htm
41. Nokia E51 Support, Available at http://www.nokia.co.in/support/product_support/nokia-e51
42. X-Lite 3.0 User Guide, Available at http://www.counterpath.com
43. ManageEngine VQManager manual, Available at http://www.manageengine.com
44. NEWT support, Available at http://research.microsoft.com
45. NetSim Brochure, Available at http://www.tetcos.com

# Chapter 5
# Technique for Improving VoIP Performance Over Wireless LANs

## 5.1 Introduction

In recent years, VoIP over WLANs has witnessed a rapid growth due to significant savings in network maintenance and operational costs and introduction of new services. While being deployed widely in enterprise and home networks, the overall performance of VoIP strongly depends on the hardware properties of the concerned access points (APs). In order to maintain liveliness or a certain degree of interactivity, real-time traffic must reach the destination within a preset time interval with some tolerance [1]. Since the VoIP-based communication is targeted to handle real-time traffic exclusively, several protocols such as RTP, cRTP, RTCP are required for the proper transport of voice packets which are highly delay sensitive. Appropriate usage of these protocols is required to maintain the QoS particularly under congested network conditions.

As VoIP communication is a real-time application, it demands minimum delay and packet loss. So far, various techniques have been proposed to maintain the QoS of VoIP sessions within acceptable limits. For example, best-effort data control and admission control policies are proposed in [2] to guarantee QoS for real-time transmissions in the IEEE 802.11e WLANs. In [3], an error protection method is devised for adaptive QoS maintenance. On the other hand, a traffic shaping scheme has been implemented in [4] for an AP while sending packets over an unstable channel. Other schemes such as in [4, 5] successfully try to avoid throughput degradation due to packet loss and retransmission on unstable links.

However, there is further scope for improvement in optimizing the concerned parameters of the APs such as selection of buffer size, RTS threshold, transmitter power, etc. Owing to frequent changes in network conditions coupled with diversity and environmental factors in addition to the unlicensed Industrial, Scientific, and Medical (ISM) band in IEEE 802.11b WLANs [6], it is difficult to develop a general algorithm for suitable optimization of AP parameters. In this chapter, this issue is resolved by developing a suitable algorithm after carefully analyzing the

© Springer International Publishing AG, part of Springer Nature 2019
T. Chakraborty et al., *VoIP Technology: Applications and Challenges,*
Springer Series in Wireless Technology, https://doi.org/10.1007/978-3-319-95594-0_5

direct and indirect factors guiding the APs and thereby suggesting an optimization technique that will help in configuring the AP parameters for optimal VoIP performance. Considering a snapshot of wireless network, close to the test-bed as far as practicable, a simulated measurement analysis is done with NetSim [7]. Results of this simulation are then applied to the developed algorithm to ascertain the optimized AP parameters. The work is further extended in optimizing the individual node parameters involved in the test-bed after rigorous analysis for both voice and video calls (which have more stringent QoS requirements).

The developed algorithm in this chapter is flexible in the sense that it can be used along with retransmission [8], link-level scheduling [4], channel error correction [3], and other optimization schemes. Implementation of active queue management policy [9] using the optimization technique has also been incorporated to support this claim.

> Owing to frequent changes in network conditions coupled with diversity and environmental factors in addition to the unlicensed Industrial, Scientific, and Medical (ISM) band in IEEE 802.11b WLANs, it is difficult to develop a general algorithm for suitable optimization of AP parameters.

## 5.2   Proposed Algorithm for Access Points

APs are an integral part of WLANs. Therefore, optimizing the performance of VoIP requires optimization of the AP parameters as well. In this section, we analyze the various factors driving IEEE 802.11b APs through extensive simulations and thereafter develop an optimization technique to configure the parameters of the AP.

### 5.2.1   Analysis

IEEE 802.11b networks are governed by several factors [6, 10]. The objective is to optimize the parameters of the APs. Primary focus is laid on the factors that affect the performance of the APs. The salient features are classified into two categories with respect to whether an AP can control it or not. The controlled factors are buffer size, retransmission limit, RTS threshold, transmission power, antenna type, location factors, and network load. Uncontrolled factors include path-loss, fading, shadowing, interference, choice of codecs.

> The objective is to optimize the parameters of the APs.

In the analysis phase, the performance of the network is observed as a whole and the AP in particular under varying network conditions. NetSim is used as the simulation tool for this purpose and accordingly, a scenario is simulated where two nodes communicate between each other in a congested environment. The parameter sets are shown in Table 5.1.

Firstly, the buffer size of the AP is varied to observe the variation in delay and loss as shown in Fig. 5.1.

It is seen that increasing the buffer size results in decrease in loss as the AP can hold more number of packets in a congested network before transmitting them. However, the delay increases with buffer size in a congested network as the AP receives out of order packets and waits for earlier packets to arrive. Then the packets are sent to the receiver in the correct order. While this mechanism reduces the jitter, the end-to-end delay increases which is highly undesirable for real-time traffic.

Next, the effect of RTS threshold parameter is studied by varying it from 0 to 2347 bytes. Any packet whose length is greater than the RTS threshold is transmitted following a RTS–CTS exchange. NetSim allows RTS–CTS mechanism up to a threshold limit of 1500 bytes. The parameters of Table 5.1 are used to study the effect of buffer size and RTS threshold. Figure 5.2 shows that delay decreases with increase in RTS threshold. A RTS threshold of 2347 bytes results in 6.04 ms of delay, whereas a RTS threshold of 0 byte results in 523.83 ms of delay. Delay increases in the latter scenario as each packet waits for a RTS–CTS exchange before getting transmitted. Loss, unlike delay is directly proportional to the RTS threshold up to certain limit depending upon the nature of congestion as shown in Fig. 5.2. As RTS–CTS mechanism aims to minimize collisions among hidden stations, increasing RTS threshold increases collisions resulting in loss up to the above limit.

Figure 5.3 shows the effect of retransmissions on the AP parameters. Increasing the retransmission limit also increases the delay and loss up to the above limit. Figure 5.4 depicts dependency of throughput of the network on retransmission limit for different path-loss exponents. The throughput starts decreasing with increasing retransmissions [11] for path-loss exponent beyond 3.5. So setting up of

**Table 5.1** Parameters in NetSim

| Parameter | Value |
|---|---|
| Buffer size | Variable |
| Retry limit | 7 |
| RTS threshold | 2347 bytes |
| Transmission type | DSSS |
| Channel | 1 (frequency: 2412 MHz) |
| Transmitter power | 100 mW |
| Channel characteristics | Fading and shadowing |
| Path-loss exponent | 3.5 |
| Fading figure | 1 (Rayleigh fading) |
| Standard deviation (shadowing) | 12 (ultra-high frequency) |

**Fig. 5.1** Variation in delay and loss with increasing buffer size

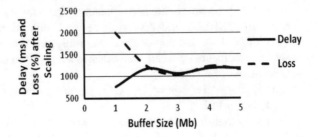

**Fig. 5.2** Variation in delay and loss with increasing RTS threshold

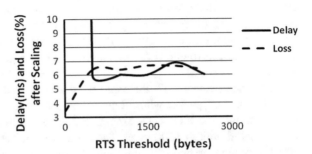

**Fig. 5.3** Increase in delay, loss with increasing retransmissions

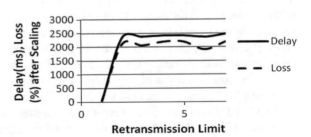

**Fig. 5.4** Variation of throughput with increasing retransmissions for various path-loss exponents

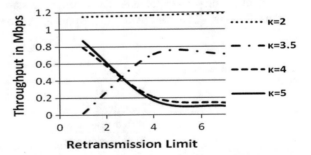

retransmission limit to a value such that both loss and delay are at an optimum level without major throughput degradation is a matter of serious concern.

Finally, the performance is analyzed under varying path-loss conditions. Rayleigh fading is considered with a standard deviation of 12 dB for shadowing as

**Fig. 5.5** Minimal transmitter power requirement with increasing distance from the AP for various path-loss exponents

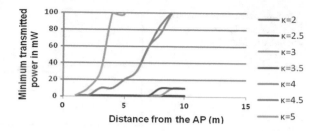

the AP operates at ultra-high frequency (UHF) of 2412 MHz [12]. The transmission power of the AP is varied to record the observations. Figure 5.5 indicates that for increasing path-loss, the minimum transmitter power required to keep the delay and loss to tolerable limits increases as nodes move away from the AP. So the AP must operate at maximum power. The aim is to increase the Equivalent Isotropically Radiated Power (EIRP) [13] which is given by (5.1).

$$\text{EIRP} = P_\text{t} \times G_\text{t} \tag{5.1}$$

where, $P_\text{t}$ denotes transmitted power and $G_\text{t}$ stands for transmitted antenna gain.

As IEEE 802.11b APs operate in the unlicensed ISM band, any nearby RF network operating at similar frequencies may cause RF interference with APs. Such interference increases with increase in transmitter power. So an optimum power must be determined. This governs the relative position of APs. As gain directly influences EIRP, a directional antenna with higher gain is advantageous than an omnidirectional antenna at the cost of more APs and greater handoff latencies.

> In the analysis phase, the performance of the network is observed as a whole and the AP in particular under varying network conditions.

### 5.2.2 Proposed Optimization Technique

#### A. Background Assumptions

Assuming that necessary error correction methods are in proper place, much importance is given to delay than packet loss particularly in light of handling real-time traffic. It is also assumed that the average number of users in the network is known in advance. The QoS metrics for ascertaining the performance of the network, namely delay, jitter, packet loss, MOS, and R-Factor, are categorized into *good, tolerable,* and *poor* limits. Beyond the *good* limit is considered as the *threshold.* Further, the path-loss exponents are measured beforehand to have an idea regarding the variation of minimum transmitter power with distance from the AP

**Fig. 5.6** Point of intersection for delay and loss curves obtained in MATLAB

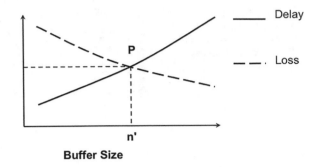

for various path-loss exponents (as shown in Fig. 5.6). Also, omnidirectional antennae are assumed to be used. This algorithm is developed with the consideration of the maximum range from the AP where the signal level deteriorates beyond the '*good*' value and delay and loss cross their respective threshold limits for VoIP sessions.

B.  Proposed Algorithm

I.  *Selection of Optimal Buffer Size (n)*

1.  Set the buffer size at the minimum level of 1 Mb.
2.  Calculate the delay and loss for the concerned buffer size.
3.  Increase the buffer size by 1.
4.  Repeat step 'b'.
5.  Repeat steps 'c' and 'd' until the buffer size reaches the maximum value of 5.
6.  Plot the two metrics against the buffer size.
7.  View trendline of the respective curves and get the relevant equations.
8.  Plot the equations in MATLAB as shown in Fig. 5.7 to find the point of intersection $P$ corresponding to the buffer size $n'$.
9.  View the $delay_p$ and $loss_p$ corresponding to the point $P$.

The intersection of the two curves may lead to two cases.

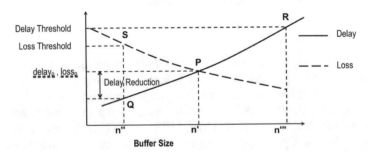

**Fig. 5.7** Delay reduction for optimal buffer size $n''$ when $delay_n$ is within the threshold value

CASE-I: Both delay$_p$ and loss$_p$ are within the *threshold* limits.

Shift the intersection point $P$ along the delay graph to the point $Q$ corresponding to the buffer size $n''$ so that the delay becomes lesser while the loss increases and reaches its threshold value. Figure 5.7 shows this shifting to utilize optimal buffer size with acceptable loss and delay.

CASE-II: Delay is unacceptable but loss is within *threshold*.

Shifting is done as in CASE-I, but the shifting is mandatory in CASE-II as shown in Fig. 5.8 to maintain the QoS of the network while the shifting is optional in CASE-I.

In both the cases, the optimal buffer size is selected as $n = n''$.

> The optimal buffer size is selected reducing the delay beyond the threshold value while the packet loss has been increased up to the safe limit.

### II. *Reduction of Packet Loss*

In the previous subsection, the optimal buffer size has been selected reducing the delay beyond the threshold value while the packet loss has been increased up to the safe limit. Now a new mechanism of RTS–CTS is being incorporated to reduce the loss beyond threshold with the optimal buffer size. The steps are described as under.

1. Plot delay and loss for selected optimal buffer size $n''$ after appropriate scaling by decreasing RTS threshold from 2347 to 0 byte.
2. View trendline of respective curves and get the relevant equations.
3. Plot the equations in MATLAB as shown in Fig. 5.9 to find the intersection point $P'$.
4. Move along delay curve from $P'$ toward $Q'$ start until delay reaches the threshold limit. Let corresponding RTS threshold be *thresh*.
5. Accept *thresh* as optimal RTS threshold if it is less than or equal to MTU in bytes.

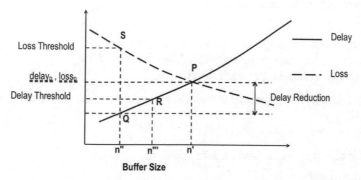

**Fig. 5.8** Delay reduction for optimal buffer size $n''$ when delay$_p$ is above the threshold value

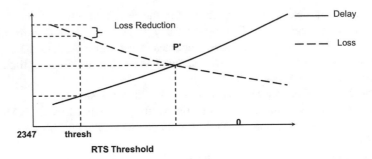

**Fig. 5.9** Loss reduction for optimal RTS threshold

A new mechanism of RTS–CTS is being incorporated to reduce the loss beyond threshold with the optimal buffer size.

III. *Selection of Retransmission Limit (r)*

1. It is applicable when both delay and loss are within the threshold limits while the path-loss exponent is less than 4.
2. Increase *r* with an initial value of 0 in steps of 1 until either the delay or loss reaches the threshold (whichever is earlier).
3. As soon as one of the metrics reach the threshold limit, the number of retransmissions (*r*) of the voice packet becomes fixed at *r'*. Thus, the limit is determined as *r'*.

IV. *Selection of Transmission Power (p) and Location (loc) of the New AP*

Let $A$ = no. of APs available, $d$ = distance of the receiver from the AP, $p$ = transmitter power, $\kappa$ = path-loss exponent, $l$ = maximum distance up to which the signal is recognizable. Accordingly, the power and location attributes are determined as per the following steps.
For $d = 1$ to $l$ {

1. Select $t_x$ from the graph of Fig. 5.5 for a given $\kappa$.
2. Measure the corresponding loss and delay from NetSim.
3. If both are within the threshold limit{

    i. If there is no adjacent RF source {Select $p$= maximum power of the relevant AP.}
    ii. Else {Select $p = t_x$.}

4. Else if *threshold* is crossed or network is out of range{

    i. If($A > 0$){

    a.  Place a new AP at location loc such that $loc = (d\text{-}10)$ m.

    b.  $A=A\text{-}1$.

    c.  Select a nonoverlapping channel.

    d.  Repeat the For loop.}

  ii.  Else{

    a.  Increase $t_x$ by external means to $t_x'$.

    b.  Select $p=t_x'$.}}}

So the developed optimization algorithm finally provides the buffer size (= $n''$), the RTS threshold value (= thresh), retransmission limit (= $r'$), transmitter power (= $p$), and location (= loc) of a new AP.

The various steps to be followed in order to find the optimum value of the above said parameters are outlined in the flowchart (Fig. 5.10).

> The developed optimization algorithm finally provides the buffer size (= $n''$), the RTS threshold value (= thresh), retransmission limit (= $r'$), transmitter power (= $p$), and location (= loc) for placing a new AP.

## 5.2.3   Implementation

The proposed algorithm is now implemented both in NetSim simulator as well as in the test-bed for improving the quality of voice calls in WLANs.

### A.  Implementation in NetSim

At first, the proposed algorithm is implemented in NetSim. A scenario is created where two nodes are communicating through voice under the presence of data traffic from other nodes present there. The scenario is implemented for path-loss exponents, $\kappa = 2$ and $\kappa = 4$.

For $\kappa = 2$ and buffer size of 1 MB, the delay and loss are 91.18 ms and 2.502%, respectively. For buffer size of 5 MB, the delay and loss become 201.32 ms and 1.232%, respectively. As per the proposed algorithm, the buffer size is considered to be 1 MB and retry limit is taken as 2. The RTS threshold limit is configured to 2347 bytes. Furthermore, the minimum transmitter power is selected to be 20 mW, and the maximum allowed location is taken to be 100 m. The voice call is made in NetSim with the parameters configured as above. The final delay is 119.66 ms, and the loss is 3.368%, both of which are within threshold limits.

For $\kappa = 4$, the delays are 150 and 493.6 ms for buffer size of 1 and 5 MB, respectively. The corresponding loss rates are 7.912 and 6.322%. Applying the proposed algorithm, the buffer size is kept at 1 MB. The retry limit of 1 is accepted, and RTS threshold limit is taken as 0 byte. The minimum transmitter power of

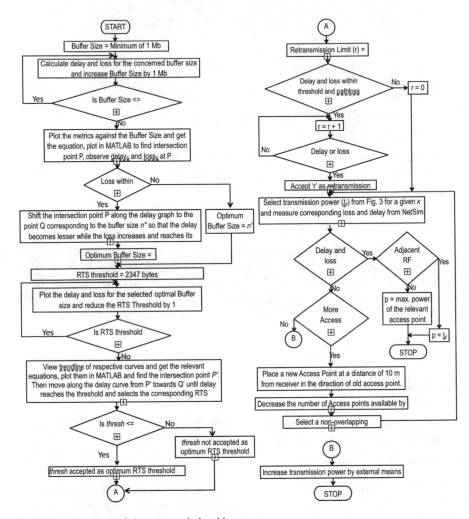

**Fig. 5.10** Flowchart of the proposed algorithm

100 mW is selected with the maximum allowed distance to be 50 m. Simulation of voice call in NetSim after configuration of the parameters results in 132.477 ms of delay and 3.277% of loss. Both of them are within tolerable limits. Thus, the proposed algorithm helps to maintain voice calls within acceptable limits of delay and loss in diverse scenarios.

> The proposed algorithm helps to maintain voice calls within acceptable limits of delay and loss in diverse scenarios.

B.  Implementation in Test-Bed

The developed optimization algorithm is further implemented in real test-bed as described in Chap. 4. Initially, a model of the test-bed is created in NetSim to obtain optimal parameter values after applying the algorithm. Then these parameters are applied in the test-bed to improve VoIP performance.

I.  *Simulation of Test-Bed Scenario*

The test-bed scenario is simulated in NetSim, and the optimization algorithm is applied. The relevant intersection point corresponds to 4.52 Mb as shown in Fig. 5.11. The corresponding delay is 158.3 ms, and the loss is 3.43%. Now as the delay crosses the threshold (150 ms) in this case but the loss is within limits, the technique is applied to select the buffer size of 2.63 Mb where the delay is 155 ms and the loss is 5% as seen in Fig. 5.12. Following the algorithm, the retransmission limit is kept to 0 and the RTS threshold is kept at 2347 bytes. As there is no other source of RF interference in our test-bed, maximum transmitter power is chosen to increase the coverage.

II.  *Calculation of Path-Loss*

The path-loss exponent plays a very important role in the proper implementation of our algorithm. So, careful measurements are taken to find out the magnitude of the path-loss in the real test-bed. The equipments used for this purpose include a mobile receiving device, i.e., a laptop, equipped with a RF power measuring software, NetStumbler. The received signal strength is measured at different distances from an AP. The collected data, as shown in Fig. 5.13, is used to find out the value of the path-loss exponent using the log normal shadowing model [14]. The model is described by (5.2).

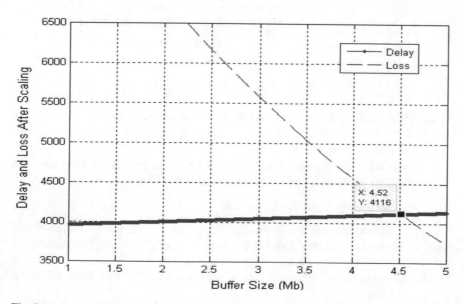

**Fig. 5.11** Intersection points of delay and loss in MATLAB

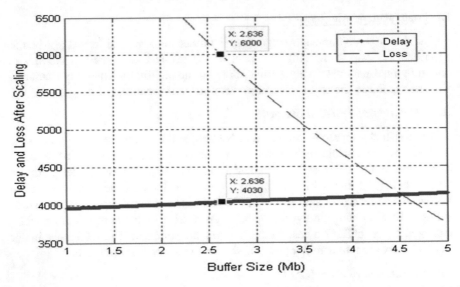

**Fig. 5.12** Optimal buffer size and the corresponding scaled delay and loss plotted in MATLAB

**Fig. 5.13** Received signal
power with respect to distance
from the AP

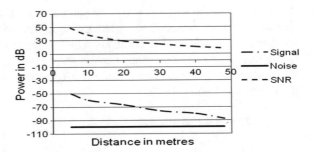

$$P_{ls}(x) \text{ (dB)} = P_l(x_0) + 10\kappa \ \log_{10}(x/x_0) + x_\sigma \qquad (5.2)$$

where, $P_{ls}(x)$ is the path-loss at a distance of $x$ meter from the AP, $P_l(x_0)$ is the
path-loss at a reference distance $x_0$ m, $\kappa$ is the path-loss exponent and $X_\sigma$ is a zero
mean Gaussian random variable.

The path-loss exponent for the test-bed scenario is calculated following the
aforementioned procedure, and its value is obtained as 3.5.

III. *Implementation in the Real Test-Bed*

All the parameters derived from the developed optimization algorithm and
calculated value of $\kappa$ are now added to the test-bed through network delay simulator
and NEWT emulator. The outcome from the test-bed is recorded through
VQManager in terms of delay and loss as shown in Fig. 5.14a, b. Both delay and

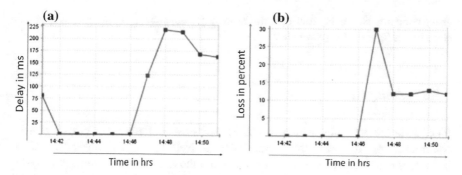

**Fig. 5.14** Optimization of delay and loss in test-bed

| Table 5.2 Implementation details | Scenario | Avg. delay (ms) | Avg. loss % | MOS |
|---|---|---|---|---|
| | Maximum queue size | 112 | 0 | 4.4 |
| | Minimum queue size | 11 | 22 | 3.1 |
| | Obtained queue size | 96 | 8 | 3.4 |

loss are minimized after initial degradation as the parameters were suitably configured following the algorithm.

Moreover, it is clearly seen from Table 5.2 that the results obtained after applying the optimization technique are optimal. Since the readings are taken at a distance of 30 m from the AP and the path-loss is calculated to be 3.5 in this test-bed, a new AP is placed at a distance of 20 m applying the algorithm. This results in further enhancement of VoIP QoS as the MOS increases to 4 from 3.4 and delay and loss decrease further.

> Both delay and loss are minimized after initial degradation as the parameters were suitably configured following the algorithm.

## 5.3 Proposed Technique for Node Parameters

The parameters of individual nodes (that act as soft phones) such as buffer size, RTS threshold, and retry limit must also be configured after proper analysis. As the number of calls increase, optimization of these parameters becomes a matter of great concern. However, this optimization must be done in conjunction with the already configured AP parameters.

### 5.3.1  Analysis

Each node has its buffer which has to be properly configured for effective voice communication. It is seen from simulations in NetSim that the buffer size of the individual nodes depends on the buffer size and the retransmission limit of the APs. Figure 5.15 shows that increasing the buffer size of the APs increases the overall delay as well. Moreover, for each AP buffer size, there is a sudden increase in delay and loss with increase in retransmission limit. So the AP buffer size must be decreased which results in further loss. Further, there has to be a minimum node buffer size for each retransmission limit to avoid considerable loss in packets as seen from Fig. 5.16. Hence, it is mandatory to carefully initialize the buffer size of the individual nodes.

> It is mandatory to carefully initialize the buffer size of the individual nodes.

The RTS threshold parameter in each node has the same function as stated in Sect. 5.3.1. Extensive simulations in NetSim point to the fact that the RTS threshold parameter of each individual node along with the RTS threshold parameter and the buffer size of the AP affects the overall delay and packet loss. Increasing the RTS threshold of the node decreases the delay significantly and increases the loss as seen in Fig. 5.17. Delay is further reduced with increase in AP RTS threshold parameter and decrease in AP buffer size. So, selecting an appropriate RTS threshold value for the individual nodes has to be done with utmost care.

**Fig. 5.15** Variation of delay with increase in retransmission limit for different buffer sizes

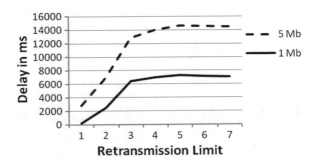

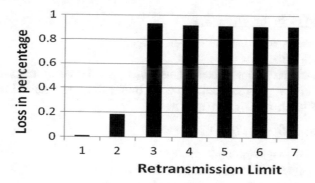

**Fig. 5.16** Variation of loss with increase in retransmission limit for buffer size of 1 Mb

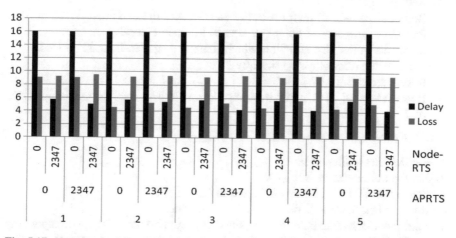

**Fig. 5.17** Variation in delay and loss for various node and AP RTS threshold parameters with increase in AP buffer size

Selecting an appropriate RTS threshold value for the individual nodes has to be done with utmost care.

## 5.3.2 Optimization Technique

A. Selection of Retransmission Limit ($r_{node}$)

For the sake of simplicity, the retransmission limit $r_{node}$ is kept the same as the retransmission limit ($r$) of the APs as obtained in Sect. 5.3.2.

B. Selection of Buffer Size ($n_{node}$)

Two scenarios are possible as given under.
SCENARIO-I: $r_{node} = 0$.
If the node has memory constraints, $n_{node}$ is selected as the maximum size available or else $n_{node}$ is selected as the AP buffer size $n$ as obtained in Sect. 5.3.2.
SCENARIO-II: $r_{node}$ is not equal to 0.

I. *Selection of Optimal Buffer Size (n)*

1. Set the retransmission limit $r$ as 1.
2. Set the buffer size at the minimum level of 1 Mb.
3. Calculate the loss for the concerned buffer size.
4. Increase the buffer size by 1.
5. Repeat step '3'.
6. Repeat steps '4', '5' until the buffer size reaches the maximum value of 5.
7. Plot the metric against the buffer size in Microsoft Excel.
8. Select the buffer size $n(r)$ after which there is a sudden increase in loss.
9. Increase the retransmission limit by 1.
10. Repeat steps from '2' to '8'.
11. Repeat steps '9', '10' until the retransmission limit reaches the maximum value of 7.

II. *Selection of the Optimal Buffer Size ($n_{node}$)*

1. Set the retransmission limit as $r_{node}$.
2. Set the buffer size as $n$ ($r_{node}$).
3. Select the AP buffer size as obtained in Sect. 5.3.2.
4. Repeat Sect. 5.3.2 B.1. for each individual node.
5. Shift the intersection point $P$ along the delay graph to the point $Q$ corresponding to the buffer size $n$ ($r_{node}$) so that the delay becomes lesser while the loss increases. This shifting is shown in Fig. 5.18. Let the delay at $Q$ be $delay_{node}$.

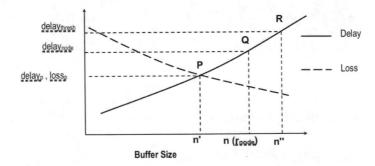

**Fig. 5.18** Selection of optimal buffer size when $n' <= n(r_{node})$ and $delay_{node} < delay_{thresh}$

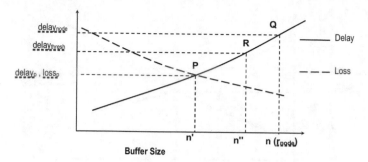

**Fig. 5.19** Selection of optimal buffer size when $n' <= n(r_{node})$ and $delay_{node} > delay_{thresh}$

6. Shift the intersection point $P$ along the delay graph to the point $R$ corresponding to the buffer size $n''$ as shown in Fig. 5.18 so that the delay reaches threshold. Let the delay at $R$ be $delay_{thresh}$.

7. Shift the intersection point $P$ along the loss graph to the point $S$ corresponding to the buffer size $n'''$ as shown in Fig. 5.21 so that the loss reaches threshold. Let the loss at $S$ be $loss_{thresh}$.

The above procedure may lead to two possible cases.

CASE-I: $n' <= n (r_{node})$

If $delay_{node} < delay_{thresh}$ the optimal buffer size is selected as $n_{node} = n (r_{node})$ as shown in Fig. 5.18. Else $n_{node} = n''$ as in Fig. 5.19.

CASE-II: $n' > n (r_{node})$

If $delay_{node} < delay_{thresh}$ the optimal buffer size is selected as $n_{node} = \min (n (r_{node}), n''')$ as seen in Fig. 5.20. Else $n_{node} = n''$ as in Fig. 5.21.

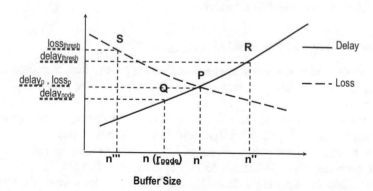

**Fig. 5.20** Selection of optimal buffer size when $n' > n(r_{node})$ and $delay_{node} < delay_{thresh}$

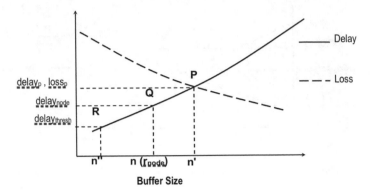

**Fig. 5.21** Selection of optimal buffer size when $n' > n(r_{node})$ and $delay_{node} < delay_{thresh}$

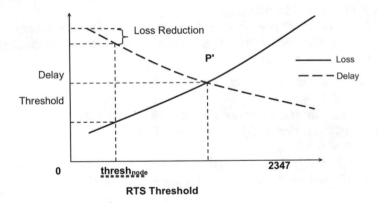

**Fig. 5.22** Selection of optimal RTS threshold

III. *Selection of RTS Threshold (thresh$_{node}$)*

1. It is applicable in scenarios where either loss is greater than threshold or both delay and loss are less than threshold. In all other cases, the default RTS of 0 is selected.
2. Plot the delay and loss for the selected optimal buffer size $n_{node}$ after appropriate scaling by increasing the RTS threshold from 0 to 2347 bytes.
3. View trendline of respective curves and get the relevant equations.
4. Plot the equations in MATLAB to find the intersection point $P'$.
5. Move along the delay curve from $P'$ toward $Q'$ until delay reaches the threshold limit. Let the corresponding RTS threshold be thresh as shown in Fig. 5.22.
6. Accept thresh as thresh$_{node}$ if it is less than or equal to MTU in bytes.

So the developed algorithm finally provides the buffer size (= $n_{node}$), the RTS threshold value (= thresh$_{node}$), and the retransmission limit (= $r_{node}$) that are to be applied in the individual nodes.

The developed algorithm finally provides the buffer size, the RTS threshold value, and the retransmission limit that are to be applied in the individual nodes.

## 5.3.3 Implementation

A. Voice Call

The proposed optimization technique is now applied in voice calls in the test-bed as described in Chap. 4. The parameters of each node are obtained after initial configuration of the AP parameters (as discussed in Sect. 5.3.2) and subsequent simulation studies of the test-bed scenario in NetSim. As soft phones are used as the nodes, there are no memory constraints. Hence, the buffer size of each node is selected to be 2.63 Mb according to the proposed algorithm. Further, the RTS threshold is configured to be 0 byte with the retry limit kept at 0 following the algorithm. The parameters obtained are now applied in a voice call in the test-bed. It is seen from Table 5.3 that loss further decreases from 8% (that was obtained through AP parameterization in Table 5.2) and is reduced to 3% when the optimization techniques are applied to both the APs and the node parameters. Moreover, the trendlines obtained through VQManager in Fig. 5.23 also imply that both delay and loss remain nearly constant thereby minimizing the chances of high jitter.

Packet loss further decreases from 8% (that was obtained through AP parameterization in Table 5.2) and is reduced to 3% when the optimization techniques are applied to both the APs and the node parameters.

**Table 5.3** Voice call details with application of optimization techniques

| Scenario | Avg. delay (ms) | Avg. loss % | MOS | Comments |
|---|---|---|---|---|
| Maximum queue size | 112 | 0 | 4.4 | Delay is high |
| Minimum queue size | 11 | 22 | 3.1 | Loss is high |
| Obtained queue size of the APs | 96 | 8 | 3.4 | MOS is acceptable but loss is >threshold |
| Obtained queue size of the individual nodes along with the obtained queue size of the APs | 82 | 3 | 4.4 | Both loss and delay <threshold and MOS is high |

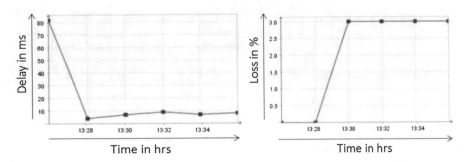

**Fig. 5.23**  Optimization of delay and loss in test-bed

## B.  Video Call

The optimization algorithm is further applied to video calls in a wireless network, which have more stringent QoS requirements in terms of delay and loss than voice calls. A snapshot of the original test-bed scenario is shown in Fig. 5.24a that will serve as the benchmark for comparison with the proposed techniques. Initially, the video calls are performed in the test-bed with the default parameters that leads to

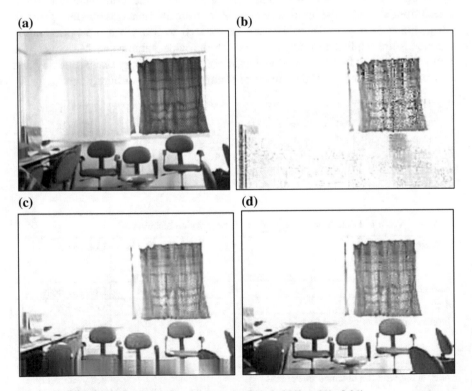

**Fig. 5.24**  Snapshots of video calls with loss of **a** 0%, **b** 22%, **c** 8%, **d** 3%

22% packet loss as observed in Fig. 5.24b. The screenshot is unrecognizable, which implies degraded nature of the video calls in the basic setup.

In the second phase, implementation of the AP optimization technique results in reduction of packet loss to 8%. A snapshot of the video call at this moment is shown in Fig. 5.24c. Here the video is almost recognizable that implies improvement in the call quality.

Finally, the optimization technique of the individual nodes is applied. A snapshot of the video call in Fig. 5.24 shows further reduction in packet loss (to 3%), and the video quality is almost the same as the default one in Fig. 5.24a. Thus, the video call is optimized in a wireless medium after application of the proposed algorithm in APs and individual nodes.

> The video call is optimized in a wireless medium after application of the proposed algorithm in APs and individual nodes.

## 5.4 Proposed Active Queue Management Policy

Queues are essential to store and forward packets following certain algorithms. A good buffer management scheme must have the following features.

a. It should maintain network in region of high throughput and low loss.
b. It should avoid bias against bursty traffic which is common in buffers with tail-drop mechanism.
c. It should avoid global synchronization. When loss occurs due to packet drop, all TCP flows reduce their window size to half and then increases slowly. This must be avoided.

Static buffers drop packets on being filled up. The packet drops are implemented with tail drop, random drop on full, and drop front on full strategies. So these buffers cannot always satisfy the aforementioned features.

Active queues drop packets before the queue is full to ensure the above features. The RED algorithm [9] is one of the active queue management policies.

It has two parts, namely

a. Estimation of average queue size

The average queue size is given by the following equation.

$$Q_{avg} = (1 - W) \times Q_{avg} + W \times Q_{sample}, \quad 0 < W < 1 \tag{5.3}$$

where

$Q_{avg}$      average queue size
$Q_{sample}$   instantaneous queue size
$W$            weight

a.  Decision of packet drop

The decision to drop the packet must depend on the average and not the instantaneous queue size. Two thresholds are defined as $Min_{th}$ and $Max_{th}$.

The decision to drop a packet as shown in Fig. 5.25 is given as follows:

- If $Min_{th} < Q_{avg} < Max_{th}$, packet is dropped with a probability '$p$'.
- If $Q_{avg} < Min_{th}$, no packet is dropped.
- If $Q_{avg} > Max_{th}$, every packet is dropped.

> Active queues drop packets before the queue is full. RED algorithm [9] is one of the active queue management policies.

The probability $p$ is given by Eq. (5.4) as follows and depicted in Fig. 5.25.

$$p = \max_p\left\{(Q_{avg} - Min_{th})/(Max_{th} - Min_{th})\right\} \tag{5.4}$$

where,

$\max_p$       maximum packet drop probability
$Q_{avg}$      average queue size
$Min_{th}$     minimum threshold
$Max_{th}$     maximum threshold

**Fig. 5.25**  Variation of packet drop probability with increasing

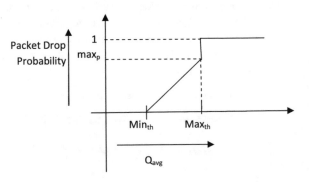

Therefore, selection of appropriate threshold limits is crucial for significant performance improvement after implementation of active queue management system. The objective is to select the optimal threshold limits, namely $Min_{th}$ and $Max_{th}$ with the help of the proposed optimization algorithm for the APs (as described in Sect. 5.3.2).

> The objective is to select the optimal threshold limits for the RED Buffer with the help of the proposed optimization algorithm for the APs.

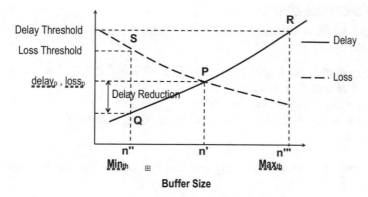

**Fig. 5.26** Selection of threshold parameters when both delay and loss are within threshold

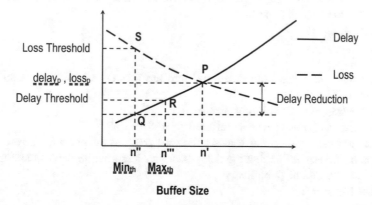

**Fig. 5.27** Selection of threshold parameters when delay is unacceptable

**Fig. 5.28** Selection of
threshold parameters when
both delay and loss are above
threshold

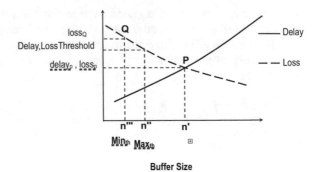

Buffer Size

B.  Optimization Technique

There are two possible cases that are considered here.
CASE-I

1. It is applicable for CASE-I and CASE-II of Sect. 5.3.2 B.1. and is shown in
   Figs. 5.26 and 5.27, respectively.
2. Select $Min_{th}$ = current buffer size $n''$.
3. Select $Max_{th}$ = buffer size corresponding to delay threshold $n'''$.

CASE-II

1. It is applicable when both delay and loss are greater than the threshold and is
   shown in Fig. 5.28.
2. Select $Max_{th}$ = current buffer size $n''$.
3. Plot delay and loss curves as referred to in Sect. 5.5.3.
4. Shift the intersection point $P$ along the loss graph to the point $Q$ corresponding
   to the buffer size $n'''$ so that the delay becomes lesser while the loss increases by
   2%. Let the loss at $Q$ be $loss_Q$.
5. Select $Min_{th}$ = $n'''$.

CASE-III

1. It is applicable when both delay and loss are greater than the threshold and is
   shown in Fig. 5.28.
2. Select $Max_{th}$ = current buffer size $n''$.
3. Plot delay and loss curves as referred to in Sect. 5.5.3.
4. Shift the intersection point $P$ along the loss graph to the point $Q$ corresponding
   to the buffer size $n'''$ so that the delay becomes lesser while the loss increases by
   2%. Let the loss at $Q$ be $loss_Q$.
5. Select $Min_{th}$ = $n'''$.

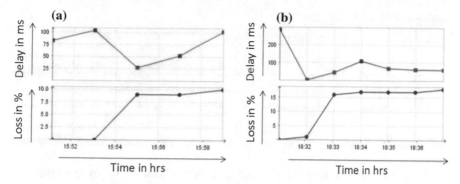

**Fig. 5.29** Variation of delay and loss for default RED threshold parameters in **a** uncongested medium and **b** congested medium

## C. Implementation

The proposed optimization technique is applied in the test-bed to select the minimum and maximum threshold values. As per the algorithm, the buffer size of 2.63 Mb is selected as the minimum threshold $Min_{th}$. The value of $Max_{th}$ is taken corresponding to the delay threshold of 180 ms. As it is more than the maximum AP buffer size of 5 Mb, $Max_{th}$ is selected as the maximum buffer size of 5 Mb.

Initially, RED is implemented with its default threshold parameters and voice calls are made under the default settings. Thereafter, RED is configured with the threshold parameters based on the obtained values, and calls are again made in this scenario. Figure 5.29a, b represents the default and the configured scenarios, respectively. It is clearly observed from the figures that under minimal network congestion, both delay and loss are comparable for RED implementations with default and configured threshold parameters, respectively.

However, the actual improvement is realized when the congestion levels increase. With increase in background traffic, loss increases when RED is implemented with default parameters as shown in Fig. 5.30a. However, RED buffers with configured parameters keep both delay and loss within tolerable limits as shown in Fig. 5.30b. This is confirmed by the delay and loss values in Table 5.4 where the

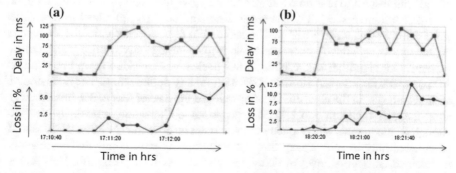

**Fig. 5.30** Variation of delay and loss for the optimized RED threshold parameters in **a** uncongested medium and **b** congested medium

**Table 5.4** RED implementation details

| Parameters | Background traffic (1 kbps) | | Background traffic (10 kbps) | |
|---|---|---|---|---|
| | Default | Configured | Default | Configured |
| Delay in ms | 66 | 61 | 70 | 63 |
| Loss in % | 6 | 6 | 14 | 4 |

configured RED buffer provides the optimal performance under congested network scenarios.

Thus, the configuration of the AP parameters also helps to configure the threshold parameters in RED which, in turn, provide an optimal performance in terms of improving call QoS.

> Configuration of AP parameters also helps to configure the threshold parameters in RED buffer which, in turn, provide an optimal performance in terms of improving call QoS.

## 5.5  Summary

In this chapter, we address the performance optimization for VoIP applications over WLAN following a three-step procedure. At first, AP parameter optimization is performed for improved VoIP performance. After examining various factors, it is concluded that for optimum performance, a certain trade-off is required among all the parameters. While this decision depends on the current network condition, a general algorithm is developed based on simulated measurements that can be applied in all such scenarios to get a fair understanding of the optimal values required for each parameter. Consequently, the optimal values of buffer size, threshold limits, transmission power, retransmission limit, etc., have been ascertained with respect to real-time traffic. Secondly, we have dealt with the problem of optimal configuration of the node parameters and devised optimization technique which further enhances VoIP performance. Finally, we configure the threshold parameters using a novel procedure while implementing active queue management systems. Our algorithm has resulted in 8% packet loss considering optimization of AP parameters in comparison with 22% without any optimization. The inclusion of node parameters in the algorithm further reduced the packet loss to 3% immediately after the call initiation process. The proposed optimization process is flexible enough to incorporate active queue management also by the selection of threshold parameters only in such a way that the RED buffers with configured parameters maintained an optimal level of delay and packet loss for the ongoing VoIP communication under congested network scenarios.

# References

1. International Telecommunication Union (ITU), *The Status of Voice over Internet Protocol (VOIP) Worldwide, 2006*. New Initiatives Programme, Document: FoV/04, 12 Jan 2007
2. Y. Xiao, H. Li, Voice and video transmissions with global data parameter control for the IEEE 802.11e enhance distributed channel access. IEEE Trans. Parallel Distrib. Syst. **15**(11), 1041–1053 (2004). https://doi.org/10.1109/TPDS.2004.72
3. Q. Li, M. Schaar, Providing adaptive QoS to layered video over wireless local area networks through real-time retry limit adaptation. IEEE Trans. Multimedia **6**(2), 278–290 (2004). https://doi.org/10.1109/TMM.2003.822792
4. M. Portoles, Z. Zhong, C. Choi, IEEE 802.11 downlink traffic shaping scheme for multi-user service enhancement, in *Proceedings of IEEE Personal, Indoor and Mobile Radio Communications (PIMRC'03)*, vol. 2, pp. 1712–1716, China, September 2003. https://doi.org/10.1109/pimrc.2003.1260407
5. P. Bhagwat, P. Bhattacharya, A. Krishna, S.K. Tripathi, Enhancing throughput over wireless LANs using channel state dependent packet scheduling, in *Proceedings of IEEE INFOCOM 1996*, vol. 3, pp. 1133–1140, USA, March 1996. https://doi.org/10.1109/infcom.1996.493057
6. Wireless LAN Medium Access Control (MAC) and Physical (PHY) Layer Specification: High Speed Physical Layer Extensions in the 2.4 GHz Band, supplement to IEEE 802.11 Standard. IEEE 802.11b, September 1999
7. NetSim Brochure. Available at http://www.tetcos.com
8. N. Kim, H. Yoon, Packet fair queueing algorithms for wireless networks with link level retransmission, in *Proceedings of IEEE Consumer Communications and Networking Conference (CCNC'04)*, pp. 122–127, USA, January 2004. https://doi.org/10.1109/ccnc.2004.1286844
9. B. Branden, et al., *Recommendations on Queue Management and Congestion Avoidance in the Internet, RFC 2309*, April 1998
10. B. Gumenyuk, Influence of different factors on wireless network connectivity, in *IEEE International Workshop on Intelligent Data Acquisition and Advanced Computing Systems* (The Czech Republic, September 2009), pp. 694–697
11. M. Heusse, F. Rousseau, G. Berger-Sabbatel, A. Duda, Performance anomaly of 802.11b, in *Proceedings of IEEE INFOCOM 2003*, vol. 6, pp. 836–843, USA, April 2003. https://doi.org/10.1109/infcom.2003.1208921
12. Wireless Channels Shadowing. Available at http://www.wirelesscommunication.nl/reference/chaptr03/shadow/shadow.htm
13. T.S. Rappaport, *Wireless Communications Principles and Practice, Pearson Education*. Asia
14. I. Saha Misra, *Wireless Communication and Networks: 3G and Beyond* (Tata McGraw Hill Education Pvt. Ltd., New Delhi, 2009), pp. 27–29

# Chapter 6
# Optimizing VoIP in WLANs Using State-Space Search

## 6.1 Introduction

VoIP is one such real-time application that is probably the most widely used on today's networks. The objective is to utilize open, flexible, and distributed implementation of PSTN-type services using IP-based signaling, routing, protocol, and interface technologies [1]. As it is being increasingly deployed in office and public networks, maintaining the QoS of an ongoing call has assumed utmost importance. Apart from optimizing the call signaling protocols, the links and nodes must also be properly tuned. As network scenarios vary over time, such optimizations must be adaptive to support a real-time communication paradigm.

> As network scenarios vary over time, such optimizations must be adaptive to support a real-time communication paradigm.

Network parameters such as available bandwidth, error rate, packet loss rate, latency vary from time to time. Each such parameter encourages the use of various ways to optimize the VoIP performance. With increasing number of users, the complexities arising out of admission control, maintenance of fairness, scalability, etc., also need to be properly addressed. So these optimization techniques must be dynamic and adaptive to the changing scenarios. However, abrupt implementation of these techniques without maintaining a proper sequence often results in degraded performance. For example, implementing active queue management in the form of RED buffer is not advantageous without any end-to-end congestion control mechanism. Further, it is observed that often such abrupt implementations of optimization techniques conflict with each other. To cite an example, RED implementation for small buffer size is not better than static queues with tail-drop mechanism [2]. However, buffer size must be kept small to reduce delay in real-time traffic under congested scenarios. So, both of these techniques tend to

© Springer International Publishing AG, part of Springer Nature 2019

T. Chakraborty et al., *VoIP Technology: Applications and Challenges*,

Springer Series in Wireless Technology, https://doi.org/10.1007/978-3-319-95594-0_6

conflict with each other. Therefore, the decision to apply the appropriate optimization technique is very crucial and must be taken carefully.

> Abrupt implementation of these techniques without maintaining a proper sequence often results in degraded performance of VoIP services.

The work in this chapter aims to generalize the problem of QoS implementation amid such diverse scenarios. The objective is to maintain adaptive QoS in multiple call scenarios and under diverse network conditions by applying the available QoS optimization techniques in a proper sequence. The focus is also on prioritizing emergency calls which must be given certain QoS guarantees. The problem of applying proper QoS mechanism to VoIP calls under varying scenarios is mapped as a state-space problem in this work. The work is extended by introducing learning in the proposed algorithm to refine the knowledge base at run-time for improvement in the call quality.

> The problem of applying proper QoS mechanism to VoIP calls under varying scenarios is mapped as a state-space problem in this work.

## 6.2   Motivation

There can be various static search strategies like depth-first search (DFS) [3], breadth-first search (BFS) [3], etc. While such search strategies can map many real-world problems, at times we need to apply heuristic-based search to reflect the dynamics of the problems. Heuristic-based search uses domain-specific knowledge to choose the successor state and therefore takes into account the varying nature of the problem environment. It is a method that may not always find the best solution but is guaranteed to find a good solution in reasonable time. Incremental search is another search strategy that solves dynamic shortest path problems, where shortest paths have to be found repeatedly as the topology of a graph or its edge costs change [4]. They can find optimal paths to similar problems more easily by reusing information from the previous problems.

Various incremental and heuristic-based search methods are being applied to solve problems in the field of symbolic planning [5, 6], path planning in the form of mobile robotics and games [7, 8], reinforcement learning [9], control problems [10], etc. Networking domain has also witnessed the usage of dynamic state-space search techniques for optimal route planning [11]. However, little work has been done with respect to mapping a particular network-related problem, other than finding shortest routes, to a state-space problem and solving it. Some of these works are discussed here. For instance, cell-to-switch assignment is developed using heuristic search for mobile calls in [12]. Heuristic search is further used in [13] to simulate possible attackers searching for attacks in modeled network for proactive and continuous identification of network attacks. Recursive random search (RRS) strategy is used in [14] to build an online simulation framework to aid generic large-scale network protocol and parameter configuration. Dynamic channel allocation in mobile cellular networks while maintaining QoS has been formulated as state-space problem in [15] and solved by applying heuristic search technique. On similar terms, this chapter aims to address the issue of QoS maintenance in real-time applications like VoIP by mapping it into a state-space search problem.

Real-time constraints on search strategies imply that search must be restricted to a small part of the domain that can be reached from the current state with a small number of action executions [16]. Real-time heuristic search is one such technique that makes planning efficient by limiting the search horizon. Korf's Learning Real-Time A* (LRTA*) method [17] is probably the most popular real-time search method used. Another technique is incremental heuristic search, which makes planning efficient by reusing information from previous planning episodes to speed up the current one as described in D* Lite [18] and Lifelong Planning A* (LPA*) [19] algorithms. Thus, real-time search techniques with an upper bound on planning and execution time are considered suitable for implementation in real-time communication domain like VoIP.

As optimizations in the field of QoS maintenance continue to evolve and mitigate the effects of unpredictable nature of networks, implementing them in proper sequence to achieve the highest performance efficiency is the biggest challenge. Moreover, each such QoS implementation mechanism must be maintained adaptively to cope up with variations in the network or changes in the user scenarios. To resolve conflicts, the system must capture system policies, including end-to-end scheduling policies, policies to decide which application's QoS to degrade when there are not enough resources to provide the desired QoS to all applications, and admissions control policies [20]. The objective is, therefore, to propose an optimization algorithm driven by real-time learning-based heuristic incremental state-space search that fulfills the aim of maintaining adaptive QoS in multiple call scenarios and under diverse network conditions by applying the available QoS optimization techniques in proper sequence. Thus, this research work is based on two principal steps, namely

1. Optimization of VoIP call using dynamic search and
2. Implementation of learning strategy for QoS enhancement.

As optimizations in the field of QoS maintenance continue to evolve and mitigate the effects of unpredictable nature of networks, implementing them in proper sequence to achieve the highest performance efficiency is the biggest challenge.

## 6.3    Why Use State-Space Approach?

A state-space search is the method of finding one or more goal states from the start state through certain intermediate states. In some problems, it is only the goal state that gives the solution, while in others, the manner in which the states are traversed to reach the goal state also becomes a part of the solution [21].

Formally, a state space can be defined as a tuple [N, A, S, G] where

- N is a set of states
- A is a set of arcs connecting the states
- S is a nonempty subset of N that contains start states
- G is a nonempty subset of N that contains the goal states.

There can be various static search strategies like depth-first search (DFS), breadth-first search (BFS). While such search strategies can map many real-world problems, at times we need to apply heuristic-based search to reflect the dynamics of the problems. In heuristic-based search, each state is given a certain heuristic or cost and traversing is done following a certain heuristic function. A heuristic is a method that may not always find the best solution but is guaranteed to find a good solution in reasonable time. It is applied to problems which could not be solved any other way or for which solutions take an infinite time or very long time to compute. Incremental search, on the other hand, reuses information from the previous searches to speed up the current search [22]. So they can find optimal paths to similar problems more easily than by solving each problem from scratch.

Incremental heuristic search combines the features of both. While it reuses information from the set of previous problems, it makes a heuristic-based search based on certain approximations to find an optimal solution in the least time. Since maintaining the QoS in each network scenario is similar with few modifications, we can apply incremental search technique. Further, as the decision to reach the goal state must be done in real time with respect to VoIP, heuristic search must be applied, which ensures search within an optimal time. So we map our problem as an incremental heuristic search problem and implement it accordingly.

> While static search strategies can map many real-world problems, at times we need to apply heuristic-based search to reflect the dynamics of the problems.

## 6.4    Optimization of VoIP Call Using Dynamic Search

The aim is to map the problem of optimizing VoIP over various network links into a state-space domain where the next state from a set of intermediate states is selected based on incremental heuristic search obeying certain constraints.

### *6.4.1   Proposed Technique*

The proposed technique designs a tuple [N, A, S, G] having four attributes that are described below.

- 'S' contains start state which is defined as call initiation state with respect to time and having heuristics, namely delay, loss, and MOS.
- 'G' contains the call termination state with respect to time along with its related heuristics as stated above.
- 'N' contains all the intermediate states within. An intermediate state is taken as any part of an ongoing call with respect to time along with its related heuristics. Heuristics can be categorized as excellent, good, average, and poor based on user satisfaction level. Any intermediate state is derived by variation in network parameters, by significant change in heuristic values, and also by the application of QoS optimization techniques.
- 'A' is a set of arcs from one state to another and is effected by transition functions. There are three transition functions, namely $\delta1$, $\delta2$ and $\delta3$. $\delta1$ is network triggered and can occur due to changes in background traffic, disconnections, network latency, loss rate, error rate, etc. $\delta2$ is performed by the user in response to $\delta1$ and involves applying various QoS optimization techniques. $\delta3$ is again user triggered, but this is applicable only in case of a multiple call scenario. It also involves maintaining the QoS of an ongoing call.
- Every heuristic must obey certain constraints. It implies that the heuristic must not cross a certain threshold value. Delay must be within 180 ms, whereas loss must be less than or equal to 5%. MOS must be more than or equal to 2. Constraints are local for each individual call. However, in multiple call scenario, global constraints are taken as mean of local constraints.

The state-space scenario for each call is shown in Fig. 6.1.

The proposed technique designs a tuple [N, A, S, G] having four attributes.

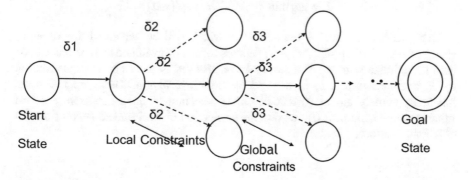

**Fig. 6.1** State-space diagram for the proposed approach

**Index**

| State-space | Significance in VoIP |
|---|---|
| Start state | Call initiation |
| Goal state | Call termination |
| Local constraints | Constraints for each call |
| Global constraints | Constraints in multiple call scenario |
| $\delta 1$ | Change in network or heuristics |
| $\delta 2$ | Optimization technique in single call |
| $\delta 3$ | Optimization technique in multiple call scenario |
|  | Best ranked action |
|  | Other actions |

Now the proposed algorithm is discussed in detail. It consists of two phases, namely analysis and implementation. Each call at a particular instant of time is taken as a state '$s$'. Each such state is associated with two metrics, namely $g = \{\text{delay, loss}\}_{\text{avg}}$ and $h = \{\text{delay}_{\text{est}}, \text{loss}_{\text{est}}\}_{\text{avg}}$. '$g$' calculates the average of the delay and loss for the states already generated, as measured by the network monitoring tool. '$h$' estimates the delay ($\text{delay}_{\text{est}}$) and loss ($\text{loss}_{\text{est}}$) for the state that is to be generated following implementation of a certain QoS mechanism. This estimation is based on simulation or in test bed and may vary from the original results.

### 6.4.1.1 Analysis Phase

This phase analyzes all possible conditions of network with respect to delay and loss. There can be four scenarios that include delay and loss within tolerable limits, worsening of either delay or loss and finally degradation of both. For each such scenario, the order of implementation of the available QoS mechanisms is selected. Each such mechanism is denoted by the action '$a$'. The order is selected based on expected performance of QoS mechanism. Mathematically, it is denoted by Eq. (1) [16].

$$f = \arg\min a \in A(s)\, h\, (\text{succ}(s, a)) \tag{1}$$

Here, successor 'succ' is the next state which will be generated due to application of action '$a$' on state '$s$'. A(s) denotes the set of available actions that can be used to optimize state '$s$'. The best ranked action '$a$' is such that by implementing it, '$h$' becomes minimum as denoted by argmin function. There are two other functions, namely one-of ($f$) that describes selection of an action from a set of suitable actions and next ($f$) that describes selection of next ranked action from set of suitable actions.

Analysis phase analyzes all possible conditions of network with respect to delay and loss. For each such scenario, the order of implementation of the available QoS mechanisms is selected.

### 6.4.1.2 Implementation Phase

As the call starts, the initial state is generated along with its heuristics. With variation in network parameters or significant change in heuristic values, new intermediate states are created along with the related heuristics. The transition function is termed as $\delta1$. Now each state is monitored to check whether the local constraints are satisfied. Each constraint has a 'threshold' value. If the local constraints are violated, $\delta2$ is applied to bring the heuristics within threshold limits. This implies that the best ranked mechanism is applied as per the analyzed results. As a new state is generated in this process, it is monitored further. If local constraints are still not satisfied, then the next best ranked action is implemented and so on.

In a multiple call scenario, the global constraints must also be satisfied. This means controlling the quality of individual call while maintaining an optimal global call scenario. The calls whose QoS metrics have degraded are classified as 'degraded' calls and the rest as 'accepted' calls. The existing QoS implementations for the accepted calls are stopped temporarily, and these QoS mechanisms are all redirected to the degraded calls. As the global constraints are satisfied, new states are generated, and the corresponding transition functions are termed as $\delta3$. All the newly generated states are again monitored, and QoS mechanisms are implemented again to satisfy the local constraints. High-priority calls may be given certain weights and in that case, the global constraints can be calculated as the weighted mean of the local constraints.

Thus, each state is associated with a heuristic and the search process for the next state is based on minimal '$h$' value. This search in turn reuses information gathered from the previous searches, and thus, it is incremental in nature.

In the implementation phase, the best ranked QoS mechanism is applied as per the analyzed results.

The pseudo-code is given as under.
Algorithm:

1. $s: = s_{start}$.    /*Call initiation state with heuristics namely delay and loss.*/
2. Calculate $g_s$.    /*The delay and loss are measured for the current state.*/

3. IF $s \in G$ THEN GOTO step 18.     /*As call ends, goal state is reached.*/
4. IF $g_s$ varies OR $g_s$ > threshold THEN GOTO step 5 ELSE GOTO step 2.
5. $s: = s_{\delta 1}$. Calculate $g_s$.     /*New state is generated due to change in network conditions or heuristics.*/
6. $a: =$ one-of $(\text{argmin}_{a \in A(s)} h(\text{succ}(s, a))$.     /*Select action that chooses successor state with minimal 'h' value.*/
7. Execute action $a$.     /*Action '$a$' is implemented.*/
8. $s: = s_{\delta 2}$.     /*New state is generated due to application of user optimization technique in the form of action '$a$'.*/
9. Calculate $g_s$.     /*The delay and loss are measured for the current state.*/
10. IF $g_s$ < threshold THEN GOTO step 13 ELSE GOTO step 11. /*Local constraints must be satisfied.*/
11. $a: =$ next$(\text{argmin}_{a \in A(s)} h(\text{succ}(s, a))$.     /*Select the next action from the set of actions that chooses successor state with minimal '$h$' value.*/
12. Execute action a. GOTO step 8.
13. IF no. of calls > 1 THEN GOTO step 14 ELSE GOTO step 3.
14. Classify ongoing calls as accepted calls whose $g_s$ < threshold. Rest of the calls are degraded calls.
15. Stop action $a \in A(s)$ for normal calls.
16. Execute action $a = \text{argmin}_{a \in A(s)} h(\text{succ}(s, a))$ for calls in degraded category.
17. $s: = s_{\delta 3}$. GOTO step 9.     /*New state is generated in multiple call scenario after application of optimization technique in the form of action '$a$'.*/
18. Calculate $g_s$ for $s \in G$.     /*Goal state heuristics are calculated.*/

This entire approach is represented pictorially in Fig. 6.2.

## 6.4.2   Implementation of the Algorithm

The proposed technique is now implemented in the test bed (which has already been described in Chap. 4). Further, we use acoustic echo cancelation, autogain control, noise reduction, and differentiated services code point (DSCP) QoS as part of maintaining signaling QoS.

### 6.4.2.1   Test-Bed Analysis Phase

Initially, the effect of buffer size in the performance of an ongoing VoIP call is studied. Buffers are used by the intermediate nodes like routers and APs as well as by the end points to store and forward packets. Four scenarios are created for analysis using NEWT [23] and network delay simulator as shown in Table 6.1. It is seen from Table 6.2 that an increase in loss rate in the network results in degraded performance in terms of loss of packets in scenario 4. As buffer size is increased, end-to-end delay increases and retransmissions take place after certain timeout,

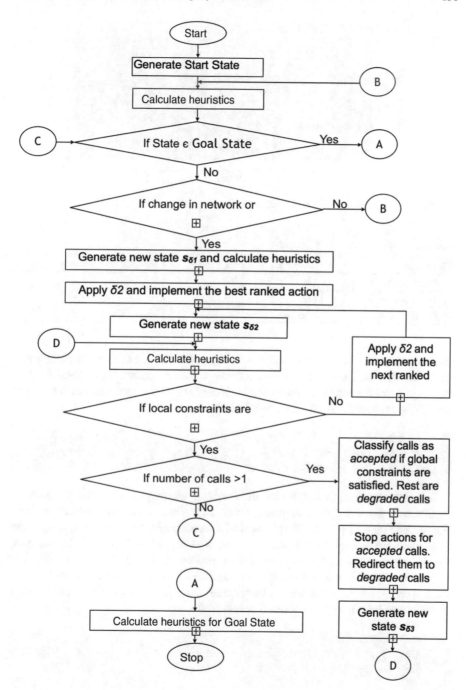

**Fig. 6.2** Flowchart depicting the proposed approach

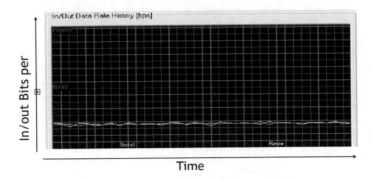

**Fig. 6.3** Effect of constant in/out bit rate on an ongoing call

**Table 6.1** Different network scenarios

| Parameters | Scenario 1 | Scenario 2 | Scenario 3 | Scenario 4 |
|---|---|---|---|---|
| Network latency in ms | 100 | 100 | 100 | 100 |
| Network loss in % | Nil | Nil | 30 | 30 |
| Buffer size in packets | Maximum | 20 | Maximum | 20 |

resulting in further loss as in scenario 3. Even in the absence of loss rate, increasing the buffer size increases the end-to-end delay while decreasing it increases the loss as observed in scenarios 1 and 2, respectively. Therefore, the selection of proper buffer size is crucial for effective VoIP performance.

Selection of proper buffer size is crucial for effective VoIP performance.

It is also observed that if the in/out bit rate in an endpoint (caller/callee) varies significantly with time, the call quality drops and terminates at last. In order to show this, BroadVoice 32 (BV32) [24] is used as the codec and the in and out buffer sizes of an endpoint are varied to obtain the results as shown in Table 6.3. In addition, Fig. 6.4 clearly depicts the call termination process as the in/out bit rate varies in contrast to Fig. 6.3 where the call continues under the effect of constant bit rate. Thus, it can be inferred that similar buffer sizes and hence comparable bit rates must be maintained for a call to continue successfully.

Similar buffer sizes and hence comparable bit rates must be maintained for a call to continue successfully.

Further, the effect of active queue management is studied. Active queues drop packets before the queue are full based on certain probabilities and threshold

**Table 6.2** Delay and loss in each scenario

| Scenarios | Max. delay (ms) | Avg. delay (ms) | Max. loss (%) | Avg. loss (%) |
|---|---|---|---|---|
| 1 | 156 | 112 | 0 | 0 |
| 2 | 39 | 11 | 44 | 5 |
| 3 | 94 | 6 | 61 | 22 |
| 4 | 6 | 1 | 49 | 21 |

**Table 6.3** Readings for the variable in/out bit rate in an endpoint

| Local → remote buffer size (packets) | Remote → local buffer size (packets) | Sending rate (kbps) | Receiving rate (kbps) | Call duration (s) |
|---|---|---|---|---|
| 20 | 20 | 26 | 26 | 1137 |
| 90 | 20 | 26 | 54 | 168 |

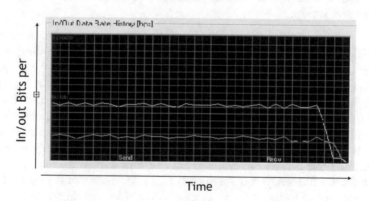

**Fig. 6.4** Effect of variable bit rate on an ongoing call

parameters. This helps them to maintain bursts in flows as well as fairness among multiple users. Here, RED [25] queue is implemented as part of active queue management. We implement RED by keeping the maximum threshold at 100 and the minimum threshold at 50. As RED is not suitable for small queues [2], it is implemented using a reasonable buffer size. We create two congested media having 1kbps and 10 kbps constant bit rate (CBR) background traffic.

As observed from Table 6.4, in a moderately congested medium, both delay and loss are within tolerable limits. Thus, RED is advantageous than having a buffer with fixed size. However, increase in congestion increases the packet loss as per the obtained readings. This is verified by the fact that static RED is not better than normal queue with tail-drop mechanism [2]. Further, the efficiency of RED implementation is not realized without any end-to-end congestion control [2] mechanism. So selecting an active queue management policy in conjunction to other QoS policies is of utmost importance toward maintaining the quality of call.

**Table 6.4** Different execution scenarios of RED implementation

| Parameters | Background traffic 1 kbps | Background traffic 10 kbps |
|---|---|---|
| Average delay in ms | 66 | 70 |
| Average loss in % | 6 | 14 |

> Selecting an active queue management policy in conjunction to other QoS policies is of utmost importance toward maintaining the quality of call.

Finally, we implement IntServ model to optimize VoIP performance. IntServ model proposes two service classes, namely

(a) Guaranteed load service for applications requiring a fixed delay bound,
(b) Controlled load service for applications requiring reliable and enhanced best-effort service.

These service classes are implemented by four components, namely flow specification, signaling protocol, admission control routine, and finally packet classifier and scheduler [26]. The controlled load service provides the requestor with service closely equivalent to that provided to uncontrolled (best-effort) traffic under lightly loaded conditions [27] and does not have specific target values for control parameters such as delay or loss. Guaranteed service guarantees that datagrams will arrive within a guaranteed delivery time [28]. As both service classes are applicable here, they are implemented in the test bed with respect to the scenarios as discussed in Table 6.1. Broadvoice-32 FEC (forward error correction) [24] is used as the codec for this analysis. Experiment results conclude that the controlled load service gives better performance in terms of packet loss than guaranteed service in scenarios 1, 2, and 3 as observed in Fig. 6.5. Further, in scenario 4 which is the worst scenario in terms of network congestion, the call terminates in an average 58 s under guaranteed service but goes on for an average 111 s under controlled load service. Thus, it can be concluded that controlled load service is more suited in highly congested scenarios than guaranteed service.

**Fig. 6.5** Effect of controlled load and guaranteed load service on packet loss in various scenarios

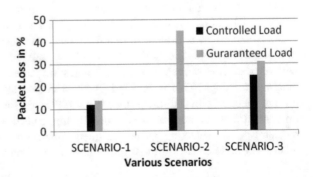

Controlled load service is more suited in highly congested scenarios than guaranteed service.

### 6.4.2.2   Selection of The Order of Implementation of Optimization Techniques

From the analyzed results, the order of implementation is proposed under varying network scenarios with respect to delay and loss. Four scenarios are considered for this phase.

Case 1: Both delay and loss are within tolerable range.
Guaranteed load service is applied to further enhance the QoS.
Case 2: Delay is tolerable but loss is high.
The buffer size of the intermediate routers or APs is increased till an acceptable value of delay. Further, RED is applied to reduce bursts in loss as the next option. If loss still persists outside the tolerable limit, the third option is to apply forward error connection (FEC) technique to recover from the loss. The last option is to apply the controlled load service.
Case 3: Loss is less but delay is high.
The buffer size of the intermediate routers or APs is decreased till an acceptable value of loss. Weighted RED is applied as the next option with drop type as random to ensure fairness. Finally, controlled load service is applied as the last option.
Case 4: Both delay and loss are poor.

Controlled load service is applied to the traffic. Further, RED is implemented with a small difference between the maximum and minimum thresholds as the next option. This helps to maintain a trade-off between delay and loss.

Thus, we suggest optimization with respect to flow classifier using DSCP, buffer size, active queue management, forward error correction, and traffic policing using various service classes, as these are the most general parameters in each network.

We suggest optimization with respect to flow classifier using DSCP, buffer size, active queue management, forward error correction, and traffic policing using various service classes, as these are the most general parameters in every network.

### 6.4.2.3   Test-Bed Implementation Phase

Our proposed approach is initially implemented in a single call scenario. The test bed as described in Chap. 4 is used for exhaustive analysis. The network conditions

are varied using NEWT and network delay simulator, and the algorithm is implemented accordingly. Figure 6.6 shows the state-space diagram for the VoIP call as
executed in the test-bed.

The heuristics for each state are shown in Table 6.5, and the transition function
for every link is described in Table 6.6. The various categories in which the
heuristics are classified are described in Table 6.7. The mean delay during the
ongoing VoIP session is 120 ms, and packet loss is 1%. The average MOS of the
overall call is 3.3. All these values suggest that the call is of an acceptable quality.
Readings from VQManager as seen in Fig. 6.7 suggest that both loss and delay are
within threshold values and remain fairly constant even as network conditions
degrade.

> Test-bed readings suggest that after application of the proposed technique,
> both loss and delay for the ongoing VoIP communication are restricted within
> threshold values and they remain fairly constant even as the network con-
> ditions degrade.

**Fig. 6.6** State transition
diagram for the call

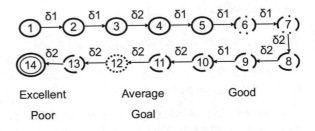

Excellent          Average          Good

Poor          Goal

**Table 6.5** Heuristics for
each state during the call

| State | Delay (ms) | Loss (%) | MOS | Duration (s) | Comments |
|---|---|---|---|---|---|
| 1 | 6 | 0 | 4.4 | 1 | Start state |
| 2 | 19 | 0 | 4.4 | 420 | Excellent |
| 3 | 85 | 0 | 4.4 | 530 | Excellent |
| 4 | 69 | 0 | 4.4 | 90 | Excellent |
| 5 | 95 | 0 | 4.4 | 350 | Excellent |
| 6 | 131 | 0 | 4.4 | 137 | Good |
| 7 | 147 | 0 | 4.4 | 126 | Good |
| 8 | 169 | 0 | 4.4 | 600 | Average |
| 9 | 164 | 0 | 4.4 | 187 | Average |
| 10 | 160 | 2 | 3.3 | 143 | Average |
| 11 | 169 | 2 | 2 | 136 | Average |
| 12 | 169 | 2 | 1.8 | 136 | Poor |
| 13 | 143 | 2 | 2 | 340 | Average |
| 14 | 163 | 2 | 2 | 250 | Goal state |

**Table 6.6** Transition function for every link between the states

| Link | $\delta 1$ | $\delta 2$ | Comments |
|------|------|------|----------|
| 1–2 | 50 ms latency | | |
| 2–3 | 65 ms latency | | |
| 3–4 | | Guaranteed service is applied | Delay decreases to some extent |
| 4–5 | 80 ms latency | | |
| 5–6 | 120 ms latency | | |
| 6–7 | | | Delay varies significantly |
| 7–8 | | Buffer size is reduced to 50 | Delay decreases with decrease in buffer size |
| 8–9 | | Buffer size is reduced to 30 | Delay has reached threshold mark |
| 9–10 | 0.01 Loss rate | | It means 1 out of every 100 packets will be lost from network |
| 10–11 | | Buffer size is increased to 45 | It is done to decrease the loss and hence enhance the MOS |
| 11–12 | | Buffer size is increased to 60 | It is done to decrease increasing loss. MOS now becomes uniform instead of decreasing |
| 12–13 | | RED is applied with max threshold of 100 and min threshold of 50 | As buffer size cannot be increased further due to increase in delay, the next best ranked action is chosen |
| 13–14 | | Controlled load service is applied | As MOS is still poor, this is done as the next best ranked one and MOS gets improved to its threshold |

**Table 6.7** Category of heuristics

| Heuristic category | Description |
|--------------------|-------------|
| Excellent | Delay $\leq$ 100 ms, loss $\leq$ 1%, MOS $\geq$ 4 |
| Good | 100 ms < delay $\leq$ 150 ms, 1% < loss $\leq$ 2%, 3.5 $\leq$ MOS < 4 |
| Average | 150 ms < delay $\leq$ 180 ms, 2% < loss $\leq$ 5%, 2 $\leq$ MOS < 3.5 |
| Poor | Delay > 180 ms, loss > 5%, MOS < 2 |

The proposed algorithm is now implemented in a multiple call scenario. Two calls are made simultaneously. While the first one is between two soft phones, the second call is made between a mobile phone and a soft phone. These calls are mapped as their state-space diagrams. The variation in call quality metrics with time is recorded in VQManager and is depicted in Fig. 6.8.

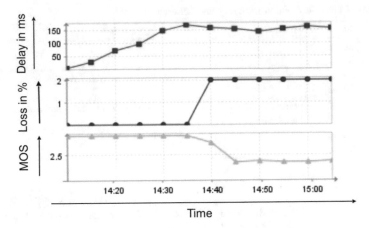

**Fig. 6.7** Variation of delay, loss, and MOS with state transitions in a single call scenario

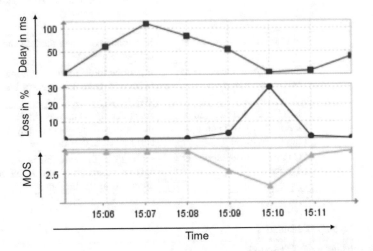

**Fig. 6.8** Variation of delay, loss, and MOS with state transitions in a multiple call scenario

It is seen from Fig. 6.8 that as overall delay of the calls increases, the buffer size is decreased as per $\delta2$ satisfying the local constraints. However, overall loss for all the calls increases. So $\delta3$ is implemented following the proposed algorithm by increasing the buffer size to satisfy the global constraints. While this increases the delay again, it is tolerable. On the other hand, loss decreases and MOS remains at a tolerable limit. The maximum, minimum, and average heuristic values are shown in Table 6.8.

After applying the global constraints, packet loss decreases, while MOS remains at a tolerable limit for all the VoIP calls.

**Fig. 6.9** Screenshots of the video call with recorded packet loss of **a** 0%. **b** 3%. **c** 4%. **d** 6%. **e** 5%

**Table 6.8** Heuristic values in multiple call scenario

| Parameters | Minimum | Maximum | Average |
|---|---|---|---|
| Delay (ms) | 4 | 110 | 49 |
| Loss (%) | 0 | 30 | 5 |
| MOS | 1.4 | 4.4 | 3.6 |

Finally, the algorithm is applied to maintain the QoS of a video call. An average video call in Internet incurs a loss of 1–2%. Network loss is therefore introduced to degrade the video call in the test-bed. Figure 6.9a, b show the quality of the received picture under 0 and 3% packet loss, respectively. As loss reaches 4%, the degradation is quite noticeable as shown in Fig. 6.9c. The picture quality slowly becomes unrecognizable as loss reaches 6% as seen from Fig. 6.9d. After implementation of the proposed algorithm, loss is again tolerable as it reaches the threshold limit of 5% and the video becomes recognizable in the degraded scenario as can be inferred from Fig. 6.9e. Thus, the proposed algorithm helps to maintain the QoS of the video call adaptively even as loss increases in the network.

> The proposed algorithm helps to maintain the QoS of the video call adaptively even as packet loss increases in the network.

## 6.5   Implementation of Learning Strategy for QoS Enhancement of VoIP Calls

It is observed from Sect. 6.4 that the QoS implementation techniques are repeated in the same order every time for each scenario. Moreover as learnt from the implementation results in Sect. 6.4.2, the best ranked mechanism may not be always suitable in a particular scenario. Hence, the next ranked action is selected from the knowledge base, and this process continues until the most suitable action is found that maintains both delay and loss within tolerable limits. Thereafter, the knowledge base is not updated, and every time, the same sequence of actions has to be repeated for reaching the acceptable call quality with respect to a particular scenario. Our objective is therefore to update the knowledge base so that the most suitable QoS implementation mechanism is applied as soon as possible. Therefore, the idea of "learning" from the networked environment is applied to the existing algorithm.

Learning is the acquisition of new declarative knowledge, the development of motor and cognitive skills through instruction or practice, the organization of new knowledge into general, effective representations, and the discovery of new facts and theories through observation and experimentation [29]. Our aim is to make the algorithm "learn" from the environment as the call goes on and accordingly update the knowledge base. The primary advantage of incorporating learning to the existing incremental heuristic search is to make the algorithm more dynamic and adaptive to the changing scenarios. However, inclusion of learning must also satisfy our initial objective as mentioned in Sect. 6.2.

> The primary advantage of incorporating learning to the existing incremental heuristic search is to make the algorithm more dynamic and adaptive to the changing scenarios.

## 6.5.1 Proposed Algorithm

The "learning" algorithm is implemented in conjunction with the previous algorithm as stated in Sect. 6.4. This algorithm comprises of two sections: (i) knowledge acquisition and (ii) skill refinement. They are described in the following subsections.

### 6.5.1.1 Knowledge Acquisition Phase

Knowledge acquisition is defined as learning new symbolic information coupled with the ability to use that information in an effective manner [29]. The essence of knowledge acquisition in this algorithm is building the knowledge base to explain broader scope of situations, to predict the behavior of the physical world and finally to be more accurate in the application of suitable actions. Initially, the knowledge base is built during the analysis phase as described in Sect. 6.4.1. The full knowledge base is finally developed with the following steps.

1. Associate each action '$a$' with its '$h$' value.
2. Rank '$a$' based on the category of heuristics in which '$h$' lies. one-of ($f$) denotes that all '$a$' have the same rank. next ($f$) denotes that '$a$' has the next lower rank.
3. Calculate $g(s)$ as '$a$' is applied to any particular state '$s$'.
4. Update $h(a) = g(s)$.

> The essence of knowledge acquisition in this algorithm is building the knowledge base to explain broader scope of situations, to predict the behavior of the physical world and finally to be more accurate in the application of suitable actions.

### 6.5.1.2 Skill Refinement Phase

Skill refinement is defined as the gradual improvement of skills by repeated practice and correction of deviations from the desired behavior [29]. Skill refinement is used in this algorithm in order to update the knowledge base in the run-time as the call goes on. Since the analysis phase as described in Sect. 6.4.1 is performed in a particular environment based on predictions and certain estimations, the knowledge base built thereafter may have certain discrepancies that are detected only in the run-time. Skill refinement phase of the 'learning' algorithm helps to eradicate these erroneous attributes of the knowledge base so that the most suitable action can be applied in the least possible time.

Let '$a_{current}$' be the final action that satisfies the local constraints after implementation on the ongoing call at the current point of time. Let $C$ contains the set of conflicting actions denoted by

   $C = \{S_1, S_2,...S_n\}$,

where $S_1 = (a_1, a_2,...,a_n)$ = one set of conflicting actions,

   $S_n = (a_{11}, a_{22},...a_{nn})$ = another set of conflicting actions, and so on. Based on this, the following steps are to be performed.


1. For all actions a of a particular scenario such that rank(a) > rank($a_{current}$)

2. {

3.                    if h(a) > h($a_{current}$)

4.                        {

5.                            if (( a and $a_{current}$) belong to $S_i$ )

6.                                {

7. rank($a_{current}$) = rank(a)

8.                                    delete a

9.                                }

10.            else

11.                                {

12.                                    swap (rank($a_{current}$) , rank(a))

13.                                }

14.}

15.    }


It is to be noted that while knowledge acquisition is the initial stage in developing the knowledge base, skill refinement focuses on updating this knowledge base so as to maintain the call quality even under varying network dynamics.


Skill refinement phase of the 'learning' algorithm helps to eradicate the erroneous attributes of the knowledge base so that the most suitable action can be applied in the least possible time.

## 6.5.2   *Implementation of the Algorithm*

The knowledge base for appropriate ranking and implementation of the best ranked action is initially developed during the analysis phase (described in Sect. 6.4.1). The knowledge base is further updated by the "learning" algorithm in this section. Both these algorithms (as described in Sects. 6.4 and 6.5, respectively) are implemented in a single call under the varying network scenarios that were previously set in Sect. 6.4. It is observed from Fig. 6.10 that the end-to-end delay decreases and the value of MOS gets enhanced with time after implementation of the algorithms.

The effect of the learning algorithm is clearly visible with respect to packet loss, as demonstrated in Fig. 6.11. In region 1, the network loss rate is increased to degrade the QoS. Actions are implemented as per their ranking in the knowledge base till the VoIP packet loss is minimized below the threshold level. This entire process has consumed approximately 10 min. In effect, the initial algorithm as described in Sect. 6.4 is implemented.

> Introduction of learning has further enhanced the call quality under both diverse and degraded scenarios.

Thereafter, the network parameters are reset and again configured to same values in region 2. With the learning algorithm in execution this time, the knowledge base is now refined. The result is quite evident in region 3 where the most suitable action is applied in the least possible time, thereby preventing any increase in VoIP packet loss unlike the previous scenario of region 1. In other words, our algorithm learns from its past experience and updates itself accordingly. On a comparative note, it is

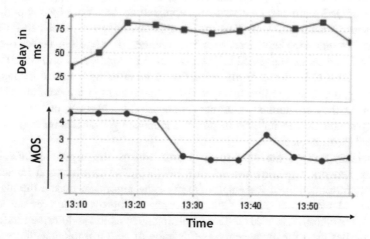

**Fig. 6.10** Variation of delay and MOS with time

**Fig. 6.11** Variation of packet loss with time

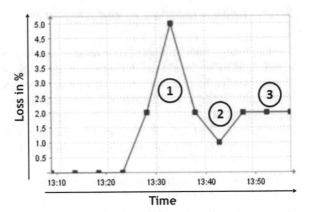

seen from Figs. 6.7 and 6.11 that while the average MOS values are comparable, the minimum value of MOS is higher in Sect. 6.5 in comparison to Sect. 6.4, thus clearly reflecting the fact that introduction of learning has further enhanced the call quality under both diverse and degraded scenarios.

## 6.6 Benefits of the Proposed Algorithm

The benefits of the proposed algorithm are discussed with respect to previous works and real-time traffic. Most of the previous state-space search approaches have been directed toward finding the optimal routes between intermediate nodes in various network conditions, such as in [30]. However, little work has been done with respect to mapping a particular network-related problem, other than finding shortest routes, to a state-space problem and solving it. Our proposed approach addresses the issue of QoS maintenance in real-time applications like VoIP by mapping it into a state-space search problem. This mapping is advantageous as it can be further optimized by applying advanced search techniques. In addition, newer optimizations, for instance, as suggested in [31], can also be applied. Current VoIP developers like CISCO implement QoS using dedicated systems for node management, link management, and traffic management. Such systems may conflict with each other unless they are applied in proper sequence. Our approach of "application of QoS implementation techniques in proper sequence" guarantees an improved and simplified performance.

Implementation of the state-space search makes the algorithm adaptive to changing network scenarios and satisfies our objective as discussed in Sect. 6.2. Similarly, introduction of learning strategy in the algorithm makes the algorithm dynamic and more intelligent. While performance improvement over the test-bed supports this claim, advanced forms of learning such as those described in [32, 33] can be further used for increasing the efficiency of action implementation.

As QoS is the main area of focus, its underlying principles (as discussed in Chap. 3) must be followed. Firstly, applications must be shielded from the complexities of the underlying QoS specification and management [34]. Our approach aims to build an automated system which will apply the transition functions to satisfy constraints. Thus, it helps in relieving the VoIP applications from the complexities of QoS maintenance, thereby maintaining transparency. Moreover, the QoS policies in our proposed algorithm are configurable, predictable, and maintainable, which are the basic aspects of QoS maintenance [34].

Viewing from the state-space search perspective, our proposed solution satisfies the following criteria.

- Inferential adequacy—Since any QoS mechanism can be fed into the transition function after proper analysis and ranking, our approach will never fall short of providing QoS under varying network scenarios.
- Inferential efficiency–Our proposed algorithm adaptively maintains the calls within acceptable delay and loss limits. While each call remains satisfactory, the global status of all the calls is also tolerable. Thus efficiency is achieved.
- Aquisitional efficiency–Existing optimization techniques can be incorporated into the proposed algorithm. For example, in our implementation, as wireless calls are being made, parameter optimizations of APs as described in Chap. 5 can also be implemented. Moreover, techniques for enhancing VoIP performance under congested scenarios can also be incorporated in this work, for example, by varying the packet payload size as suggested in Chap. 9. Advanced artificial intelligence search techniques can be further applied.

The proposed approach of "application of QoS implementation techniques in proper sequence" guarantees an improved and simplified performance. Moreover, the QoS policies in our proposed algorithm are configurable, predictable, and maintainable, which are the basic aspects of QoS maintenance.

## 6.7   Summary

In this chapter, we have dealt with the problem of adaptive QoS maintenance under dynamic and diverse network conditions and applied optimization technique in the form of incremental heuristic search. Test-bed readings verify the fact that the application of the proposed algorithm in single and multiple voice and video calls maintains both delay and loss within threshold limits even as network conditions vary with respect to time. The algorithm further ensures that no conflict arises during the application of QoS mechanisms as proper sequence is maintained among them. Performance improvement is observed after the introduction of learning strategy as it refines the knowledge base and makes the algorithm more efficient in terms of application of the most suitable QoS implementation technique in the least possible time under diverse network scenarios. While VoIP traffic binds this algorithm to real-time heuristic search, modern optimizations in this dynamic search domain can be further applied to the state-space search approach.

# References

1. International Telecommunication Union (ITU), The status of voice over internet protocol (VOIP) worldwide, 2006, New Initiatives Programme, Document: FoV/04, 12 Jan 2007
2. M. May, J. Bolot, C. Diot, B. Lyles, Reasons not to deploy RED, in *Proceedings of Seventh International Workshop on Quality of Service*, pp. 260–262, London, (1999)
3. S. Russell, P. Norvig, *Artificial Intelligence: A Modern Approach*, 3rd edn. (Pearson, New York, 2014)
4. G. Ramalingam, T. Reps, An incremental algorithm for a generalization of the shortest-path problem. J. Algorithm **21**(2) (1996)
5. B. Bonet, H. Geffner, Heuristic search planner 2.0. Artif. Intell. Mag. **22**(3), 77–80 (2000)
6. S. Koenig, D. Furcy, C. Bauer, Heuristic search based replanning, in *Proceedings of the International Conference on Artificial Intelligence Planning and Scheduling*, pp. 294–301 (2002)
7. S. Koenig, C. Tovey, Y. Smirnov, Performance bounds for planning in unknown terrain. Artif Intell **147**, 253–279 (2003)
8. L. Romero, E. Morales, E. Sucar, An exploration and navigation approach for indoor mobile robots considering sensor's perceptual limitations, in *Proceedings of the International Conference on Robotics and Automation*, pp. 3092–3097 (2001)
9. S. Koenig, M. Likhachev, Incremental A*, in *Advances in Neural Information Processing Systems*, pp. 1539–1546 (2002)
10. Al-Ansari, Efficient Reinforcement Learning in Continuous Environments, Ph.D. Dissertation, College of Computer Science, Northeastern University, Boston (Massachusetts), 2001
11. T. Zhu, W. Xiang, Towards optimized routing approach for dynamic shortest path selection in traffic networks, in *International Conference on Advanced Computer Theory and Engineering*, pp. 543–547, 20–22 (2008)
12. S. Mandal, D. Saha, A. Mahanti, Heuristic search techniques for cell to switch assignment in location area planning for cellular networks, in *IEEE International Conference on Communications*, vol. 7, pp. 4307–4311, 20–24 (2004)
13. V. Franqueira, Finding multi-step attacks in computer networks using heuristic search and mobile ambients, Ph.D. Dissertation, University of Twente, Netherlands (2009)
14. Y. Tao, H.T. Kaur, S. Kalyanaraman, M. Yuksel, Large-scale network parameter configuration using an on-line simulation framework. IEEE/ACM Trans. Netw. **16**(4), 777–790 (2008). https://doi.org/10.1109/TNET.2008.2001729
15. S. Mandal, D. Saha, An efficient technique for dynamic channel allocation (DCA) in mobile cellular networks, in *IEEE International Conference on Personal Wireless Communications*, ICPWC 2005, pp. 470–473, 23–25 (2005). https://doi.org/10.1109/icpwc.2005.1431390
16. S. Koenig, Real-time heuristic search: research issues, in *International Conference on Artificial Intelligence Planning Systems,* Pennsylvania (1998)
17. R. Korf, Real-time heuristic search. Artif. Intell. **42**(2–3), 189–211 (1990)

18. S. Koenig, M. Likhachev, D* Lite, in *Proceedings of the National Conference on Artificial Intelligence*, pp. 476–483 (2002)
19. S. Koenig, M. Likhachev, D. Furcy, Lifelong planning A*. Artif. Intell. J. **155**(1–2), 93–146 (2004)
20. S. Chatterjee, J. Sydir, B. Sabata, T. Lawrence, Modeling applications for adaptive QoS based resource management, in *Proceedings of the 2nd IEEE High Assurance Systems Engineering Workshop*, USA (1997)
21. E. Rich, K. Knight, *Artificial Intelligence*, 2nd ed. (McGraw-Hill, New York, 1990)
22. S. Koenig, M. Likhachev, Y. Liu, D. Furcy, Incremental heuristic search in artificial intelligence. AI Mag. **25**(2), 99–112 (2004)
23. NEWT Support. Available at http://research.microsoft.com
24. J-H. Chen, W. Lee, J. Thyssen, RTP payload format for broadvoice speech codecs, RFC 4298 (2005)
25. B. Branden, et. al., *Recommendations on Queue Management and Congestion Avoidance in the Internet*, RFC 2309 (1998)
26. B. Li, M. Hamdi, D. Lang, X. Cao, Y.T. Hou, QoS-enabled voice support in the next-generation internet: issues, existing approaches and challenges. IEEE Commun. Mag. **38** (4), 54–61 (2000)
27. J. Wroclawski, *Specification of the Controlled-Load Network Element Service*, IETF RFC 2211 (1997)
28. S. Shenker, C. Partridge, R. Guerin, *Specification of Guaranteed Quality of Service*, RFC 2212 (1997)
29. R.S. Michalski, J.G. Carbonell, T.M. Mitchell, *Machine Learning: An Artificial Intelligence Approach* (M. Kaufmann, Klein Vielen, 1986)
30. T. Zhu, W. Xiang, Towards optimized routing approach for dynamic shortest path selection in traffic networks, in *International Conference on Advanced Computer Theory and Engineering*, pp. 543–547, Thailand, 20–22 Dec 2008
31. V. Bulitko, N. Sturtevant, J. Lu, T. Yau, Graph abstraction in real-time heuristic search. J. Artif. Intell. Res. **30**(2), 51–100 (2007)
32. S.R. Sutton, A.G. Barto, *Reinforcement Learning: An Introduction* (MIT Press, Cambridge, 1998)
33. J. Baxter, A model of inductive bias learning. J. Artif. Intell. Res. **12**, 149–198 (2000)
34. C. Aurrecoechea, A.T. Campbell, L. Hauw, *A survey of QoS architectures, Multimedia Systems*, vol. 6, Issue 3 (Springer, Berlin, 1998), pp. 138–151

# Chapter 7
# Optimization of Codec Parameters to Reduce Packet Loss Over WLAN

## 7.1 Introduction

Commercial deployment of VoIP services over WLANs cannot be achieved unless a majority of the VoIP QoS issues is solved efficiently. The 802.11 standard [1], as most of the IP networks and their underlying technologies, was not created having voice in mind and, therefore, severely limits the successful deployment of VoIP. Voice transmission over the wireless link under variable channel conditions can easily suffer from an increased packet error and loss ratio, with direct effect on its performance and quality. While solutions have been proposed in terms of efficient admission control and QoS implementation mechanisms besides protocol enhancements, codec adaptation algorithms have also seen the light of the day. In a VoIP scenario, this could be done by changing the voice codecs of one or several calls to new ones with lower overall bandwidth requirements.

> While solutions have been proposed in terms of efficient admission control and QoS implementation mechanisms besides protocol enhancements, codec adaptation algorithms have also seen the light of the day.

Every codec adaptation algorithm can be considered to follow a decision policy, which dictates the usage of voice codecs and the manner in which that the adaptation is performed, or in other words, if and how many calls should change codec, depending on the criterion to be maximized [2]. There is a clear trade-off between the different criteria, especially between the achieved quality and the quantity of accepted calls and depends on the user and system requirements.

Various works as suggested in [3, 4] have focused on switching of codecs such that more calls can be accepted at a tolerable limit. However, few works have ventured into the domain of AP management with respect to codec parameters. This chapter aims to reduce the bottleneck in APs with decrease in the codec rate

© Springer International Publishing AG, part of Springer Nature 2019
T. Chakraborty et al., *VoIP Technology: Applications and Challenges*,
Springer Series in Wireless Technology, https://doi.org/10.1007/978-3-319-95594-0_7

(measured in packets per second or pps in short) as a number of calls increase in the network. While this reduces the packet loss due to buffer flow, further implementation of active queues in conjunction to this mechanism ensures that timely action is taken to change the codec bit rate as buffers start getting filled up.

Few works have ventured into the domain of AP management with respect to codec parameters.

## 7.2  VoIP Codecs

### 7.2.1  Overview

A codec stands for coder–decoder that converts an audio signal into compressed digital form for transmission and then back into an uncompressed audio signal for replay. This conversion is accomplished by sampling the audio signal several thousand times per second. It converts each tiny sample into digitized data and compresses it for transmission. Advanced algorithms are applied by codecs to sample, sort, compress, and packetize audio data. The CS-ACELP algorithm [5] (CS-ACELP = conjugate-structure algebraic-code-excited linear prediction) is one of the most prevalent algorithms in VoIP.

The principle is to apply a determined standardized algorithm to an uncompressed digital audio and reduce number of bits required to represent that audio. Choice of codecs and the related algorithms must be initially determined before VoIP communication starts. Both parties participating in the phone call must agree on the codec using the process called "capabilities exchange" that is normally being implemented by the underlying call signaling protocol, for example Session Initiation Protocol (SIP). Every VoIP soft phone supports a list of VoIP codecs covering a range of performance levels and, often, bandwidths. During the call establishment (or session initiation) phase, they share their list of supported codecs. When more than one common codec is supported by either party, the involved parties implement a policy to prioritize a certain codec over others based on certain parameters like better audio bandwidth, lower bit rate.

### 7.2.2  Codec Parameters

There are some important codec parameters that guide their overall performance. Codec bit rate [6] is the most important one. Based on the codec, this is the number of bits per second that need to be transmitted to deliver a voice call and is given by the Eq. (1) [6].

$$\ldots\ldots\text{codec bit rate} = \text{codec sample size} \div \text{codec sample interval} \quad (1)$$

Codec sample size is the number of bytes captured by the DSP at each codec sample interval, and codec sample interval is the sample interval at which the codec operates. MOS is another important parameter and a system of grading the voice quality of telephone connections. With MOS, a wide range of listeners judge the quality of a voice sample on a scale of one (bad) to five (excellent). The scores are averaged to provide the MOS for the codec. The voice payload size is the parameter that represents the number of bytes (or bits) that are filled into a packet. It must be a multiple of the codec sample size. For example, G.729 packets can use 10, 20, 30, 40, 50, or 60 bytes of voice payload size. The voice payload size can also be represented in terms of the codec samples. For example, a G.729 voice payload size of 20 ms (two 10 ms codec samples) represents a voice payload of 20 bytes [(20 bytes × 8)/(20 ms) = 8 kbps].

Pps [6] is the primary performance metric. It represents the number of packets that need to be transmitted every second in order to deliver the codec bit rate and is given by the Eq. (2) [6].

$$\text{pps} = \text{codec bit rate} \div \text{voice payload size} \quad (2)$$

One of the most important factors to consider while building packet voice networks is proper capacity planning. Within capacity planning, bandwidth calculation is an important factor to consider when one designs and troubleshoots packet voice networks for good voice quality.

The total bandwidth is given by Eq. (3) [6].

$$\text{band width} = \text{total packet size} \times \text{pps} \quad (3)$$

where total packet size = (L2 header: MP or FRF.12 or Ethernet) + (IP/UDP/RTP header) + (voice payload size).

### 7.2.3   Evaluation of Codec

Each conferencing environment has its own acoustical challenges that require an appropriately designed conferencing solution. Accordingly, codecs must be selected after careful evaluation of the several factors [7] that are discussed as follows.

1. Audio Bandwidth: Audio bandwidth is synonymous with audio fidelity that is defined as the ability to carry audio ranging from very low pitches to very high pitches. Therefore, more bandwidth is better. Today, more codecs are available that

support 7 kHz audio because 7 kHz itself provides an easily achievable and dramatic improvement in voice-only communications.

**Inference: The more the better**

2. Data Rate: Data rate or bit rate plays a prominent role in determining the QoS of VoIP transmission especially in bandwidth-limited network conditions and while supporting a large number of simultaneous VoIP transmissions. It must be clearly understood that bit rates do not increase with rising audio bandwidth but depend on the selected codec. Codecs use compression algorithms that take different approaches toward compressing the audio samples based on the consideration whether multiple talkers or music is included or not.

**Inference: The fewer the better**

The data rates along with the audio bandwidth for some popular codecs are shown in Table 7.1.

3. Audio Quality Loss: Audio quality is a subjective measure that has more to do with user satisfaction than the quantitative metrics associated with a VoIP call. Therefore, it is different from packet loss which is a measurable quantity and an indication of the current network congestion. Subjective tests in the form of MOS can be applied in such scenarios to evaluate the performance of codecs. However, it must be ensured that all the codecs are evaluated under identical testing conditions and at their default bit rate values. In this regard, a lot depends on whether the codec is designed to handle only human speech or a combination of speech, music, and other sources of audio.

**Inference: The fewer the better**

4. Processor Requirements: Codec complexity becomes a determining factor in ascertaining the minimum requirements of the underlying processors. High-complexity codecs require relatively more faster and expensive processors along with higher memory. Further, the requirements multiply especially when VoIP phones perform multiparty bridged calls internally, which is a common VoIP-enabled feature nowadays. The processor requirements for different codecs are enlisted in Table 7.2.

**Table 7.1** Bandwidth versus bit rate for some popular codecs

| Bandwidth (kHz) | Typical bit rate (kbps) for popular codecs |
|---|---|
| 3.3 | 8 (G.729), 56 (G.711) |
| 7 | 10 (G.722.2), 24 (G.722.1), 64 (G.722) |
| 14 | 32 (G.722.1C) |
| 20 | 32 (G.719), 64 (AAC-LD) |
| 22 | 32 (Siren22) |

**Table 7.2** MIPS versus audio bandwidth for some popular codecs

| Bandwidth (kHz) | Codec | MIPS |
|---|---|---|
| 7 | G.722.2 | 38 |
| 7 | G.722 | 14 |
| 7 | G.722.1 | 5.5 |
| 20 | G.719 | 18 |
| 20 | MPEG-4 AAC-LD | 36 |

**Inference: The fewer the better**

5. Latency: Codec Latency is total amount of delay incurred by the compression algorithm implemented by the codec. Increased complexity increases the overall latency which is not at all desirable with respect to VoIP (considering the maximum latency limit of 150 ms that VoIP transmission can tolerate = 150 ms). Moreover, jitter buffers which are used to reduce variable delay components in VoIP packets cause a significant amount of latency. These are sometimes embedded within a codec, which makes it important to ensure that multiple, redundant jitter buffers are not inadvertently built into a system.

**Inference: The fewer the better**

6. Cost: Cost plays a driving factor in selecting a particular codec to be implemented in a VoIP communication. Some codecs are launched with license fees or royalties, with fees charged on a per-year or per-port or per-phone basis. The logic for paid service is that commercial VoIP providers have to incur long and expensive research for the production of efficient codecs. However, there are also royalty-free codecs like G.722 and Speex, either because the underlying patent has expired or to help the overall VoIP community at no cost.

**Inference: The lower the better**

7. Standardization and Availability: Every codec must be standardized to ensure its reliability and widespread acceptance by VoIP service providers. The ITU is the de facto worldwide agency for standardization of telecommunications codecs. Hence every standardized codec has successfully passed open, rigorous multivendor evaluation, which increases its credibility. Proprietary codecs may be incorporated in local, private VoIP deployments, but in business VoIP telephony systems, ITU-T-approved codecs are mandatory for worldwide interoperability and high reliability. Moreover, ITU-T approval guarantees availability of codecs to all service providers on fair and reasonable terms.

**Inference: The standardization the better**

## 7.2.4   Codec Compression Techniques

The most significant aspect of a codec is to efficiently compress the VoIP samples before sending them through the network in the shape of IP packets. Generally, two basic variations of 64 kbps PCM are used for these purposes, namely μ-law and a-law. Their behavior is almost similar in the sense that both these methods use logarithmic compressions to achieve 12–13 bits of linear PCM quality in 8 bits. However, they differ in their usage with respect to different countries. While North America prefers μ-law technique, Europe and other countries tend to follow μ-law. It must be noted in this regard that if one wishes to make a long distance call in between the countries opting different modulations, any required μ-law to a-law conversion is the responsibility of the μ-law country. Latest research developments have concluded that out of the two, μ-law is more advantageous as it provides better low-level signal-to-noise ratio performance.

Few codecs also exploit the redundant characteristics of the waveform itself, also referred to as waveform codecs. For example, ITU-T G.726 uses adaptive differential pulse code modulation (ADPCM), where linear prediction is used to encode the differences in amplitude and their rates of change, rather than directly encoding the speech amplitude themselves. Both PCM and ADPCM are examples of waveform codecs. In general, these compression techniques include variations such as linear predictive coding (LPC), code-excited linear prediction compression (CELP), and multipulse, multilevel quantization (MP-MLQ) and accordingly vary in their default MOS values. The ITU-T scoring of MOS values for some commonly used codecs is shown in Table 7.3.

A list of commonly used codecs in the VoIP domain along with their standardizations and parameters is listed in Table 7.4. However, it must be noted that codec development is a continuous process, and performance efficiency of a codec also relies upon how it cooperates with other activities in the VoIP domain.

The overall bandwidth consumed by the VoIP traffic depends on selecting the various parameters of VoIP codecs, as enlisted in Table 7.5. Again, this is a very crucial aspect with respect to capacity planning in the network.

**Table 7.3**  ITU-T MOS score for different codecs based on their compression methods [8]

| Compression method | Bit rate (kbps) | Sample size (ms) | MOS score |
|---|---|---|---|
| G.711 PCM | 64 | 0.125 | 4.1 |
| G.726 ADPCM | 32 | 0.125 | 3.85 |
| G.728 low-delay code-excited linear predictive (LD-CELP) | 15 | 0.625 | 3.61 |
| G.729 conjugate-structure algebraic-code-excited linear predictive (CS-ACELP) | 8 | 10 | 3.92 |
| G.729a CS-ACELP | 8 | 10 | 3.7 |
| G.723.1 MP-MLQ | 6.3 | 30 | 3.9 |
| G.723.1 ACELP | 5.3 | 30 | 3.65 |
| iLBC Freeware | 15.2 | 20 | 3.9 |

**Table 7.4** Standard VoIP codecs and their parameters

| Name | Standardized by | Description | Bit rate (kb/s) | Sampling rate (khz) | Frame size (ms) | Remarks |
|---|---|---|---|---|---|---|
| (ADPCM) DVI (http://www.cs.columbia.edu/~hgs/audio/dvi.html) | Intel, IMA (http://www.cs.columbia.edu/~hgs/audio/dvi/) | ADPCM | 32 | 8 | Sample | |
| G.711 (http://www.cs.columbia.edu/~hgs/audio/g711.html) | ITU-T (http://www.itu.int/rec/recommendation.asp?type=folders&lang=e&parent=T-REC-G.711) | Pulse code modulation (PCM) | 64 | 8 | Sample | μ-law (US, Japan) and A-law (Europe) companding |
| G.721 | ITU-T (http://www.itu.int/rec/recommendation.asp?type=folders&lang=e&parent=T-REC-G.721) | Adaptive differential pulse code modulation (ADPCM) | 32 | 8 | Sample | Now described in G.726, obsolete |
| G.722 (http://www.cs.columbia.edu/~hgs/audio/g722.html) | ITU-T (http://www.itu.int/rec/recommendation.asp?type=folders&lang=e&parent=T-REC-G.722) | 7 kHz audio-coding within 64 kbit/s | 64 | 16 | Sample | Subband codec that divides 16 kHz band into two subbands, each coded using ADPCM |
| G.722.1 (http://www.cs.columbia.edu/~hgs/audio/g722.1.html) | ITU-T (http://www.itu.int/rec/T-REC-G.722.1/e) | Coding at 24 and 32 kbit/s for hands-free operation in systems with low frame loss | 24/32 | 16 | 20 | See also (http://www.picturetel.com/siren/siren.htm) |

(continued)

**Table 7.4** (continued)

| Name | Standardized by | Description | Bit rate (kb/s) | Sampling rate (khz) | Frame size (ms) | Remarks |
|---|---|---|---|---|---|---|
| G.723 | ITU-T (http://www.itu.int/rec/recommendation.asp?type=folders&lang=e&parent=T-REC-G.723) | Extensions of recommendation G.721 adaptive differential pulse code modulation to 24 and 40 kbit/s for digital circuit multiplication equipment application | 24/40 | 8 | Sample | Superceded by G.726, obsolete. This is a completely different codec than G.723.1 |
| G.723.1 (http://www.cs.columbia.edu/~hgs/audio/g723.1.html) | ITU-T (http://www.itu.int/rec/T-REC-G.723.1/e) | Dual rate speech coder for multimedia communications transmitting at 5.3 and 6.3 kbit/s | 5.6/6.3 | 8 | 30 | Part of H.324 video conferencing. DSP Group (http://www.dspg.com/prodtech/truespch/6353.htm). It encodes speech or other audio signals in frames using linear predictive analysis-by-synthesis coding. The excitation signal for the high rate coder is multipulse maximum likelihood quantization (MP-MLQ) and for the low rate coder is algebraic-code-excited linear prediction (ACELP) |
| G.726 (http://www.cs.columbia.edu/~hgs/audio/g726.html) | ITU-T (http://www.itu.int/rec/T-REC-G.726/e) | 40, 32, 24, 16 kbit/s adaptive differential pulse code modulation (ADPCM) | 16/24/32/40 | 8 | Sample | ADPCM, replaces G.721 and G.723 |
| G.727 | ITU-T (http://www.itu.int/rec/T-REC-G.727/e) | 5-, 4-, 3- and 2-bit/sample embedded adaptive differential pulse code modulation (ADPCM) | Var. | ? | Sample | ADPCM. Related to G.726 |

(continued)

**Table 7.4** (continued)

| Name | Standardized by | Description | Bit rate (kb/s) | Sampling rate (khz) | Frame size (ms) | Remarks |
|---|---|---|---|---|---|---|
| G.728 (http://www.cs.columbia.edu/~hgs/audio/g728.html) | ITU-T (http://www.itu.int/rec/T-REC-G.728/e) | Coding of speech at 16 kbit/s using low-delay code-excited linear prediction | 16 | 8 | | CELP. Annex J offers variable bit rate operation for DCME |
| G.729 (http://www.cs.columbia.edu/~hgs/audio/g729.html) | ITU-T (http://www.itu.int/rec/T-REC-G.729/e) | Coding of speech at 8 kbit/s using conjugate-structure algebraic-code-excited linear prediction (CS-ACELP) | 8 | 8 | 10 | Low delay (15 ms) |
| GSM 06.10 (http://www.cs.columbia.edu/~hgs/audio/gsm.html) | ETSI | Regular-pulse excitation long-term predictor (RPE-LTP) | 13 | 8 | 22.5 | Used for GSM cellular telephony |
| LPC10e (FIPS 1015) (http://www.cs.columbia.edu/~hgs/audio/lpc10e.html) | US Govt. | Linear predictive codec | 2.4 | 8 | 22.5 | Ten coefficients |

**Table 7.5** Bandwidth requirements and essential codec parameters

| Codec information | | | | Bandwidth calculations | | | | | |
|---|---|---|---|---|---|---|---|---|---|
| Codec and bit rate (kbps) | Codec sample size (Bytes) | Codec sample interval (ms) | Mean Opinion Score (MOS) | Voice payload size (Bytes) | Voice payload size (ms) | Packets per second (PPS) | Bandwidth MP or FRF.12 (kbps) | Bandwidth w/ cRTP MP or FRF.12 (kbps) | Bandwidth Ethernet (kbps) |
| G.711 (64 kbps) | 80 | 10 | 4.1 | 160 | 20 | 50 | 82.8 | 67.6 | 87.2 |
| G.729 (8 kbps) | 10 | 10 | 3.92 | 20 | 20 | 50 | 26.8 | 11.6 | 31.2 |
| G.723.1 (6.3 kbps) | 24 | 30 | 3.9 | 24 | 30 | 33.3 | 18.9 | 8.8 | 21.9 |
| G.723.1 (5.3 kbps) | 20 | 30 | 3.8 | 20 | 30 | 33.3 | 17.9 | 7.7 | 20.8 |
| G.726 (32 kbps) | 20 | 5 | 3.85 | 80 | 20 | 50 | 50.8 | 35.6 | 55.2 |
| G.726 (24 kbps) | 15 | 5 | | 60 | 20 | 50 | 42.8 | 27.6 | 47.2 |
| G.728 (16 kbps) | 10 | 5 | 3.61 | 60 | 30 | 33.3 | 28.5 | 18.4 | 31.5 |
| G722_64 k (64 kbps) | 80 | 10 | 4.13 | 160 | 20 | 50 | 82.8 | 67.6 | 87.2 |
| ilbc_mode_20 (15.2 kbps) | 38 | 20 | NA | 38 | 20 | 50 | 34.0 | 18.8 | 38.4 |
| ilbc_mode_30 (13.33 kbps) | 50 | 30 | NA | 50 | 30 | 33.3 | 25.867 | 15.73 | 28.8 |

## 7.3 Motivation

Codecs therefore play an important role in VoIP, and hence, the dream of achieving perfect call quality in VoIP cannot be fulfilled without addressing the issues related to codecs. In multirate WLANs, users can suffer transmission rate changes due to the link adaptation mechanism. This results in a variable capacity channel, which is very hostile for VoIP and can cause serious Quality of Service (QoS) degradation in all active calls. Various codec adaptation mechanisms have been proposed as a solution to this, as well as to solve congestion problems on WLAN environments [2]. However, most of them follow reactive strategy based on packet loss to switch codecs. In this chapter, a special codec parameter, namely pps, has been analyzed and an algorithm based on proactive strategy is proposed to decrease the pps in APs to reduce the packet loss due to buffer overflow.

Adapting parameters of VoIP flows (like codec or packetization interval used) has been an area of active research and has witnessed considerable development. Evaluation of congestion due to rate change is analyzed in [9], and codec change is proposed for the node suffering rate change. In [10], when the channel is detected congested (based on real-time control protocol (RTCP) packet loss and delay feedback information), a central element performs codec adaptation using common transcoding methods for calls entering wireless cell. Focus shifts from multirate effects to congestion provoked by existence of additional VoIP sessions on the cell in [11] where the algorithm focuses on adapting low-priority calls in favor of high-priority ones by "down-switching" codec and packetization interval of low-priority calls based on a set of information.

Our objective is to decrease pps without significant latency and hence conserve more bandwidth in congested scenarios. Moreover, in WLANs as a number of users increase, there is the condition of bottleneck in the APs. Therefore, increasing pps will further add to the loss due to buffer overflow which must also be reduced. While this refers to either decreasing the bit rate or increasing the voice payload size, the pros and cons of each such procedure must be carefully analyzed before proposing any suitable algorithm.

## 7.4 Analysis

The aim is to analyze packet loss and delay due to buffer overflow with increase in pps. Therefore, the scenario is chosen where an AP manages only six communicating nodes to avoid packet loss due to network congestion. NetSim [12] is used as the simulator for this analysis as it helps to achieve the objective mentioned beforehand. Further, it is possible to map the test-bed scenario in NetSim that helps us in implementing the results obtained from this analysis into the real test-bed.

**Table 7.6** Codec parameters

| Codecs | Rate (kbps) | Voice payload size (bytes) | pps |
|--------|-------------|----------------------------|------|
| G.711 | 64 | 160 | 50 |
| G.723 | 6.3 | 24 | 33.3 |
| G.729 | 8 | 20 | 50 |
| GSM-FR | 13.2 | 33 | 50 |

Initially, standard codecs are chosen with different pps and payload sizes, and they are analyzed in the simulator. Table 7.6 shows the parameters for the chosen codecs.

Figure 7.1 shows that loss increases even in an uncongested scenario with increase in pps. As the number of packets arriving at the AP increases, each packet has to further contend for the wireless medium following a RTS–CTS exchange. Therefore, the buffer gets filled up quickly leading to loss due to overflow. The loss is high for codecs G.711, G.729, and GSM-FR with high pps. G.711 accounts for the highest loss among them as its voice payload size is the maximum.

Moreover, it is also observed that G.711 incurs the highest medium access delay of 15.445 ms compared to other codecs that experience a delay of around 5 ms. This is because of the high bit rate as well as the voice payload size in G.711. Given the significant effects of queuing delay and transmission delay on the total delay, a medium access delay of more than 10 ms in an uncongested scenario is not tolerable. The situation will be even more worse in a congested scenario when every packet will have to contend for the busy medium, coupled with the loss of RTS–CTS packets that will enhance the delay and loss further.

Thereafter, the codec parameters are varied to achieve a reduction in loss and delay. While the bit rate is kept fixed at 64 kbps, the voice payload size is varied along with the inter-arrival time. In effect, pps is decreased as voice payload size is increased while keeping the bit rate constant. The parameters are listed in Table 7.7.

It is seen from Fig. 7.2 that an increase in the voice payload size results in increase in loss even in an uncongested scenario. This is due to buffer overflow in

**Fig. 7.1** Variation in packet loss for standard codecs

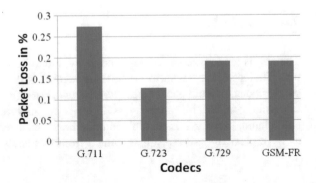

**Table 7.7**  Variation of codec parameters

| Scenarios | Rate (kbps) | Voice payload size (bytes) | Inter-arrival time (microsec) | PPS |
|---|---|---|---|---|
| 1 | 64 | 50 | 6250 | 160 |
| 2 | 64 | 60 | 7500 | 133 |
| 3 | 64 | 80 | 10,000 | 100 |
| 4 | 64 | 100 | 12,500 | 80 |

**Fig. 7.2** Variation of packet loss with increase in packet payload size

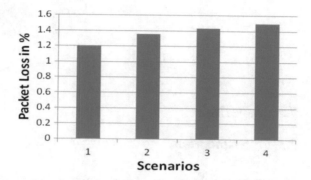

the APs as they get filled up quickly. At the same time, as observed in Fig. 7.3, the medium access delay also increases with increase in payload size as more packets are above the RTS threshold.

Finally, it is observed from Fig. 7.4 that as pps decreases, the queuing delay also decreases. This is because less packets are being generated as the payload size is increased resulting in a sharp decline in delay due to waiting in the queue. It is also observed that the total delay in the uncongested medium is largely influenced by the queuing delay, and hence, the total delay also decreases as pps decreases.

Moreover, as discussed in Chap. 5, having an extended buffer size to decrease loss in packets is not a solution as it results in higher latency that is undesirable. Hence, an optimum buffer size must be considered instead of extending it further.

**Fig. 7.3** Variation of medium access delay with increase in packet payload size

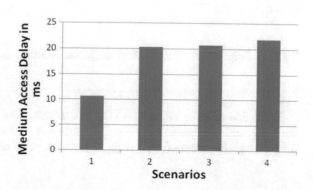

**Fig. 7.4** Variation in total
and queuing delay with
increase in packet payload
size

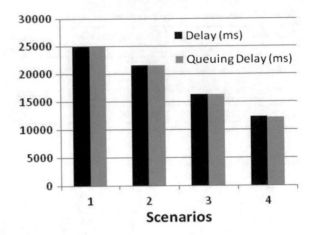

## 7.5   Algorithm

### 7.5.1   Assumptions

The switching to lower and higher bit rate codecs is done based on the sender and receiver reports that are created from network loss and delay. Therefore, it is assumed that the soft phone can switch to such rates based on the reports. The QoS metrics for ascertaining the performance of the network, namely delay, jitter, packet loss, MOS, and R-factor, are categorized into good, tolerable, and poor limits. Beyond the good limit is considered as the threshold denoted by thresh.

### 7.5.2   Algorithm

The algorithm is comprised of two methods, namely implementation of active queue management and adaptive variation of bit rates. Both these complement each other and function simultaneously. A pictorial representation of the algorithm is presented in Fig. 7.5.

#### 7.5.2.1   Implementation of Active Queue Management

1. *Calculation of queue occupancy factor*

We define queue occupancy (Q.O.) as the factor that guides the configuration of the RED buffer parameters in APs. Q.O. indicates how much the buffer is occupied for a certain time interval on a scale of 1–10. A value of more than 5 indicates that

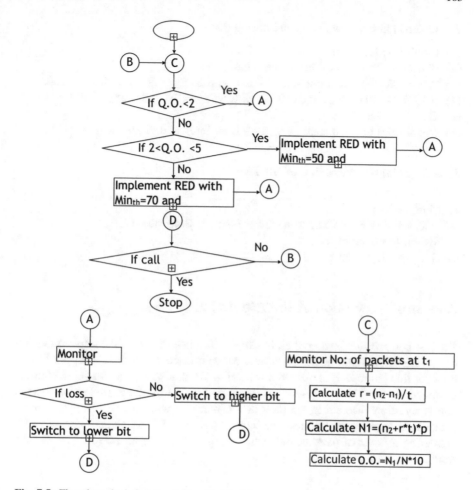

**Fig. 7.5** Flowchart depicting the proposed algorithm

the queue is likely to get filled up soon. It is an approximation and may vary from
the actual result. Let the total queue size be $N$. It is calculated as follows.

(1) Monitor the number of packets that are currently in the queue at instant $t_1$. Let it
be $n_1$.
(2) Monitor the number of packets that are currently in the queue at instant $t_2$. Let it
be $n_2$. The time interval $t = t_2 - t_1$ is set beforehand.
(3) The rate of occupancy $r$ is calculated as $r = (n_2 - n_1)/t$. A negative value
suggests decrease in number of packets in the queue.
(4) Let $N_1 = (n_2 + r*t)*p$ where $p$ = packet size
(5) Q.O. $= N_1/N*10$

2. Implementation of active queue management

(1) Monitor Q.O. factor.
(2) If Q.O. < 2 then goto the second phase.
(3) If 2 < Q.O. < 5 then implement RED with $Min_{th}$ = 50 and $Max_{th}$ = 100.
(4) If Q.O. > 5 then implement RED with $Min_{th}$ = 70 and $Max_{th}$ = 100.
(5) Goto the second phase.
(6) If call does not terminate, wait for a time interval and goto step 1.

### 7.5.2.2  Adaptive Variation of Bit Rates

1. Monitor loss
2. If loss > thresh then switch to lower bit rate. Goto step 4
3. Switch to higher bit rate.
4. Goto step 6 of Phase 1.

## 7.6  Implementation of the Algorithm

The test-bed scenario as described in Chap. 4 has been used for the implementation of the algorithm. Initially, a high bit rate codec is selected. Wideband Speex [13] is used for this purpose. It is observed from Fig. 7.6 that both loss and delay increase as pps increases. This is because as buffers tend to get filled up, packet loss occurs due to overflow. Moreover, queuing delay increases with increase in pps as more packets contend for the wireless medium. As RTS–CTS exchange takes place for increasing number of packets, delay rises further. The overall MOS degrades as a result.

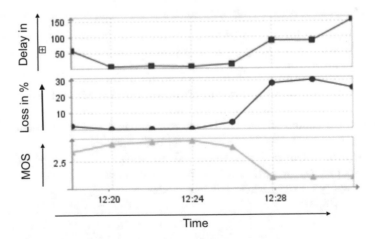

**Fig. 7.6** Variation in delay, loss, and MOS with time in high bit rate scenario

Next, the second phase of the proposed algorithm is only implemented. This refers to adaptive switching to lower bit rate while fixed buffer is implemented in the APs. Speex variable bit rate (VBR) codec [13] is used for this purpose. Figure 7.7 shows that both loss and delay are reduced as adaptive switching to lower bit rate is done. The overall MOS increases as well.

Finally, we implement RED [14] along with adaptive switching of codec bit rate. This refers to the proposed algorithm as described in Sect. 7.5.2. As observed in Fig. 7.8, both loss and delay get reduced here as well.

However, they increase quite early than in the previous scenario. This is because RED ensures that switching to lower bit rate is done even while the queue is not

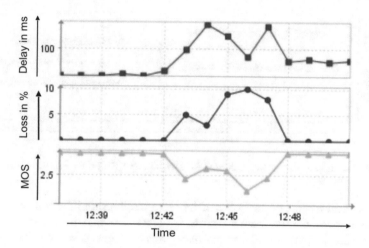

**Fig. 7.7** Variation in delay, loss, and MOS with time in low bit rate fixed buffer scenario

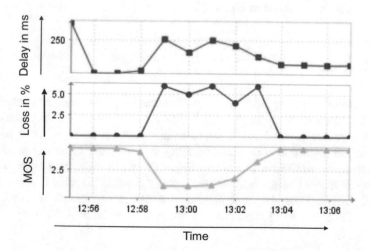

**Fig. 7.8** Variation in delay, loss, and MOS with time in low bit rate RED buffer scenario

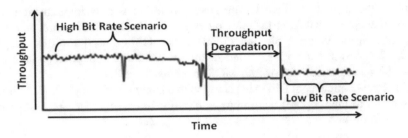

**Fig. 7.9** Variation in throughput with time for low bit rate fixed buffer scenario

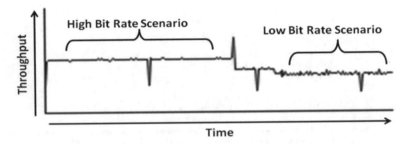

**Fig. 7.10** Variation in throughput with time for low bit rate RED buffer scenario

full. This ensures none of the QoS metrics cross the threshold limits as it is seen in the previous case. Earlier, when switching is done only when the queue is full, even after decrease in pps, loss decreases only when the queue starts getting empty. Figure 7.9 shows throughput degradation as this switching is done. However, RED implementation ensures throughput decreases with switching to lower bit rate but does not degrade as shown in Fig. 7.10.

## 7.7   Summary

In this chapter, we have addressed the problem of packet loss and increase in latency in APs due to increase in pps. After careful analysis, we observe that increase in packet payload size to decrease pps may not be an effective solution. Thereafter, an algorithm is proposed that suggests decrease in bit rate to decrease pps coupled with RED implementation. Experimental readings verify that both loss and delay reduce without significant throughput degradation after implementation of the algorithm. Thus, the configuration of codec parameters plays an important role in ensuring efficient performance of the APs in voice over WLANs.

# References

1. IEEE Standards, Available at http://standards.ieee.org/about/get/802/802.11.html
2. A. Sfairopoulou, B. Bellalta, C. Macian, M. Oliver, A comparative survey of adaptive codec solutions for VoIP over multirate WLANs: a capacity versus quality performance trade-off, EURASIP J. Wireless Commun. Netw. **534520** (2011). https://doi.org/10.1155/2011/534520
3. M.F. Tüysüz, H.A. Mantar, A cross layer QoS algorithm to improve wireless link throughput and voice quality over multi-rate WLANs, in *Proceedings of the 6th International Wireless Communications and Mobile Computing Conference*, pp. 209–213, France (2010)
4. A. Sfairopoulou, B. Bellalta, C. Macian, How to tune VoIP codec selection in WLANs? IEEE Commun. Lett. **12**(8), 551–553 (2008)
5. Rec. ITU-T G.729, *Coding of speech at 8 kbit/s using conjugate structure algebraic-code-excited linear-prediction* (CS-ACELP) (1996)
6. Voice Over IP—Per Call Bandwidth Consumption Manual. Available at http://www.cisco.com/en/US/tech/tk652/tk698/technologies_tech_note09186a0080094ae2.shtml
7. J. Rodman, White Paper on "VoIP to 20 kHz: Codec Choices for High Definition Voice Telephony", Polycom (2008)
8. M. Bhatia, J. Davidson, S. Kalidindi, S. Mukherjee, J. Peters, *Voice Compression*, Book Chapter (CISCO Press, Indianapolis, 2006)
9. P. McGovern, S. Murphy, L. Murphy, Addressing the link adaptation problem for VoWLAN using codec adaptation, in *Proceedings of the Global Telecommunications Conference (GLOBECOM '06)*, USA (2006)
10. A. Trad, Q. Ni, H. Afifi, Adaptive VoIP transmission over heterogeneous wired/wireless networks" in *Interactive Multimedia and Next Generation Networks*, vol. 3311 of Lecture Notes in Computer Science, pp. 25–36 (2004)
11. B. Tebbani, K. Haddadou, Codec-based adaptive QoS control for VoWLAN with differentiated services, in *Proceedings of the 1st IFIPWireless Days (WD '08)*, France (2008)
12. NetSim Brochure. Available at http://www.tetcos.com
13. Speex Information. Available at http://www.speex.org/
14. B. Branden, et. al., *Recommendations on queue management and congestion avoidance in the internet*, RFC 2309 (1998)

# Chapter 8
# QoS Enhancement Using an Adaptive Jitter Buffer Algorithm with Variable Window Size

## 8.1 Introduction

The transmission of real-time audio and video applications over the Internet is a challenging task. These applications have strict bounds in terms of delay and loss. This set of applications is called real-time applications and they include facilities like Internet protocol (IP) telephony, teleconference. Of the various real-time applications, we have concentrated on Voice over IP (VoIP) since it has gained importance over the past few years owing to its low cost and ease of interfacing between data and voice traffic [1].

However, the characteristics of the Internet backbone are time-variant in nature. Its properties are so random and unpredictable that it is not an easy task to statistically determine the way the backbone is going to behave in a future point of time. The reason behind the lack of proper prediction of its characteristics is its dependency on the behavior of the other connections throughout the network [2]. The connectivity may be hampered due to several reasons rendering networking applications ineffectual. The networks suffer from congestion when traffics exceeding the capacity of the network are routed through it. As a result, the data packets suffer high delay and loss while passing through the network. Such delay and loss are unacceptable when we want to utilize the network for transmitting voice traffic.

The unpredictability of the network characteristics may also cause the consecutive packets passing through the network to suffer from different extents of end-to-end delay. Generally, the VoIP applications send data at a constant rate. So, any alteration in the end-to-end delay suffered by two voice packets means that the time between two events occurring at the source and the time between the events perceived at the receiver are not equal. Such an event is not desired in a VoIP system as it degrades speech quality. This variation in delay suffered by two consecutive voice packets is termed as inter-arrival jitter [1, 3]. VoIP applications are not only sensitive to the extent of the delay and loss suffered by the voice

© Springer International Publishing AG, part of Springer Nature 2019
T. Chakraborty et al., *VoIP Technology: Applications and Challenges*,
Springer Series in Wireless Technology, https://doi.org/10.1007/978-3-319-95594-0_8

packets but also to the inter-arrival jitter. Jitter is one of the main factors that degrade the QoS in IP networks [4]. Hence, we need some mechanisms to undo this variation in delay that is being incorporated into the voice packets by the network.

> The variation in delay suffered by two consecutive voice packets is termed as inter-arrival jitter

One of the feasible solutions is the use of some mechanisms which aim to reduce the network congestion, e.g., increasing the packet payload size [5]. A set of standard voice coder–decoders (codecs) of different bit rates can also be employed. In this method, a lower bit rate codec is used when the network becomes congested so as to reduce the traffic introduced to the network. A mechanism using a Fast-Startup Transmission Control Protocol (TCP) has been explored in [6] in order to find a possible way to enhance VoIP performance in congested networks. Another method that seeks to reduce congestion with the help of distributed routing has been proposed in [7, 8]. However, when the network conditions become favorable for transmission, higher bit rate codec is used to achieve better QoS [8]. Congestion can also occur in the intermediate access points. So, VoIP performance can also be enhanced by optimizing the access point parameters [9].

The most effective solution to minimize jitter is to store the voice packets for a short time in the receiver buffer before playing it out, thus reducing the jitter [10]. However, using a fixed playout time for every packet is rendered useless if the network characteristics are variable and the voice packets suffer different extent of delay while passing through it [11]. Several algorithms have already been proposed so that the playout time for a voice packet is delayed in accordance with the variation in the network and thus provide better QoS for the VoIP applications. However, if the playout delay is too large, the end-to-end delay suffered by the voice packets is increased beyond acceptable limits and the VoIP performance may become irritating to the user because significant awkwardness occurs between speakers when delay exceeds 200 ms [11]. On the other hand, if the playout delay is too small, voice packets may be discarded due to late arrival [1]. This leads to loss in information and hampers the voice quality. It is not desirable that the voice packets suffer from either high delay or high loss. So, it is mandatory to obtain an optimum playout time to get the best performance out of a VoIP system.

> So, it is mandatory to obtain an optimum playout time to get the best performance out of a VoIP system.

In this chapter, we have analyzed some of the already established adaptive jitter buffer playout algorithms and have tested for their efficiency in several network scenarios. Further, we have also taken a note of their shortcomings and have proposed a new adaptive jitter buffer playout algorithm that provides the optimum QoS to the VoIP application in terms of delay, loss, and jitter. The performance of the new algorithm has been tested in varying network scenarios using OPNET Modeler 14.5.A. The effect of varying the sliding window size on the performance of the proposed algorithm has also been tested. Moreover, its performance has also been compared with the analyzed algorithms.

## 8.2   Background Study

The Internet mainly relies on IP for proper routing of the packets passing through it. So it is clear that in the network layer, IP is the life force of a VoIP system. Both Transmission Control Protocol (TCP) and User Datagram Protocol (UDP) can be used in the transport layer. However, owing to its lower bandwidth requirements, UDP is preferred over TCP [1, 3]. However, UDP lacks the reliability provided by TCP. So, another protocol, the Real-time Transport Protocol (RTP), rides on top of UDP and IP to ensure that the voice packets are able to meet their stringent time requirements [12]. RTP provides time stamps and sequence numbers to voice packets. The receiver can use the sequence number to determine whether the packets have arrived in order and the time stamps to assess the inter-arrival jitter suffered by the packets en route to the VoIP receiver. Real-time Transport Control Protocol (RTCP) works in tandem with RTP. It provides control information to the VoIP source and the VoIP receiver, allowing them to work efficiently. The primary function of RTCP is to provide feedback on the quality of data distribution [12].

In a VoIP system, inter-arrival jitter is an estimate of the statistical variance of the voice packet inter-arrival time. The inter-arrival jitter is defined to be the mean deviation of the difference in packet spacing at the receiver as compared to the sender for a pair of packets. The inter-arrival jitter is given by (8.1) [12].

$$J(i) = J(i-1) + \{|D(i-1), i| - J(i-1)\}/16 \qquad (8.1)$$

where $J(i)$ is the jitter for the $i$th voice packet, and for two packets $i$ and $j$, $D(i, j)$ is given by (8.2) [12].

$$D(i, j) = (R_j - R_i) - (S_j - S_i) = (R_j - S_j) - (R_i - S_i) \qquad (8.2)$$

where $S_i$ and $S_j$ are the RTP time stamps for packets $i$ and $j$, and $R_i$ and $R_j$ are the time of arrivals in RTP time stamp units for packets $i$ and $j$.

It is evident from (8.2) that the inter-arrival jitter can be easily calculated from the delay suffered by two consecutive voice packets while traveling from the VoIP source to the VoIP receiver.

In a VoIP system, inter-arrival jitter is an estimate of the statistical variance of the voice packet inter-arrival time.

Figure 8.1 illustrates that the time required for both packet A and packet B to travel from the source to the receiver are same. However, packet C encounters delay while traveling through the network and arrives after its expected time. The reverse case may also happen, where a packet may be routed quicker due to sudden increase in the network capacity and the packet reaches its destination before its expected time of arrival. This is the reason that we require a jitter buffer to smooth out the inter-arrival jitter. It may be noted that for non-real-time systems, packets can be stored indefinitely at the local buffer. However, for real-time applications like VoIP, delayed packets may become useless after a certain point of time. The jitter buffer holds the delayed packets in an attempt to neutralize the effects of packet inter-arrival jitter. This helps maintain the liveliness of real-time communication over packet-switched networks. The jitter buffer must be neither too small nor too large. If the jitter buffer is too small, it will not serve its purpose; on the other hand, if it is too large, it may remain filled with useless packets and hence waste memory space [3].

> The jitter buffer holds the delayed packets in an attempt to neutralize the effects of packet inter-arrival jitter.

The need to evaluate the voice quality is important for evaluating the performance of various jitter buffer algorithms. Accordingly, metrics as already defined in Chap. 3 such as MOS and R-factor are used for this purpose.

The E-Model is most commonly used for objective measurements. The basic result of the E-Model is the calculation of the R-factor. The R-factor is defined in terms of several parameters associated with a voice channel across a mixed switched circuit network and a packet-switched network. The parameters included in the computation of the R-factor are fairly extensive covering such factors as echo, background noise, signal loss, codec impairments, and others. R-factor can be expressed by (8.3) [13].

$$R = 94.2 - I_d - I_e \qquad (8.3)$$

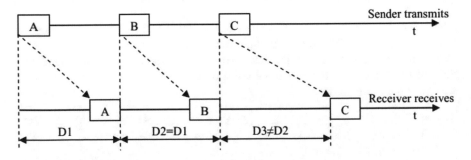

**Fig. 8.1** Time diagram showing jitter

where $I_d$ is the impairment associated with the mouth-to-ear delay of the path, and $I_e$ is an equipment impairment factor associated with the losses within the gateway codecs. $I_d$ is given by (8.4).

$$I_d = 0.024\, T_d + 0.11(T_d - 177.3)\, H(T_d - 177.3) \qquad (8.4)$$

where $H(x) = 0$ if $x < 0$

$$H(x) = 1 \text{ if } x \geq 0$$

$T_d$ is the end-to-end delay suffered by the voice packet en route from the VoIP source to the VoIP receiver. $I_e$ is given by (8.5).

$$I_e = A \times \ln(1 + B \times \text{loss ratio}) + C \qquad (8.5)$$

where A, B, and C are constants depending on the codec used.

MOS is related to $R$-factor by (8.6) [13].

For $R < 0$: MOS $= 1$
For $R > 100$: MOS $= 4.5$
For $0 < R < 100$: MOS $= 1 + 0.035R + 7 \times 10^{-6}R(R - 60)(100 - R)$    (8.6)

## 8.3  Related Work

A significant research has already been conducted in the quest of finding a suitable adaptive jitter buffer playout algorithm. As already mentioned, finding the proper playout delay is of utmost importance. In order to find out the efficiency of some of the already existing adaptive jitter buffer algorithms, we have studied five algorithms mentioned in [14, 15].

### 8.3.1  Exponential Average Algorithm (EXP-AVG) [14]

In this algorithm, the delay estimate for the $i$th packet is computed based on RFC 793 algorithm [16] and the variation in the delays is calculated as suggested by Van Jacobson in the calculation of round-trip-time estimates for the TCP retransmit timer [17]. In this algorithm, the estimate of the playout delay for packet $i$ is evaluated by the Eq. (8.7).

$$P_i = d_i + 4v_i \qquad (8.7)$$

where $d_i$ and $v_i$ can be figured out by (8.8) and (8.9).

$$d_i = \alpha d_i - 1 + (1 - \alpha)n_i \qquad\qquad (8.8)$$

$$v_i = \alpha v_i - 1 + (1 - \alpha)d_i - n_i \qquad\qquad (8.9)$$

where $n_i$ denotes the one-way delay of the $i$th packet and the value of $\alpha$ is $v_i = v_i - 1 + (1 - \alpha)d_i - n_i$ 0.998002 [14].

### 8.3.2   Fast Exponential Average Algorithm (F-EXP-AVG) [14]

This algorithm is similar to the previous one. The only difference being that if the current packet's network delay '$n_i$' is greater than $d_{i-1}$, then $d_i$ is given by Eq. (8.10).

$$d_i = \beta d_i - 1 + (1 - \beta)n_i \qquad\qquad (8.10)$$

where the value of $\beta$ is 0.75 [14].

### 8.3.3   Minimum Delay Algorithm (Min-D) [14]

The primary objective of this algorithm is to minimize the delay. So, it uses the minimum value of the network delay suffered by the packets in the current talkspurt to estimate the playout delay of the next talkspurt. Let $S_i$ be the set of all packets received in the talkspurt prior to the one initiated by $i$. So, the delay estimate for packet $i$ is calculated by (8.11). Apart from this modification, this algorithm is similar to the EXP-AVG algorithm.

$$d_i = \min j \in S_i\{n_i\} \qquad\qquad (8.11)$$

where $\beta$ is 0.75 [14].

### 8.3.4   Spike Detection Algorithm (Spike-Det) [14]

One of the most common phenomena that can be observed in a VoIP system is that some of the packets may suddenly suffer from high end-to-end delay. The result of this effect is that no voice packet reaches the receiver for some time followed by the arrival of a large number of voice packet reaching almost simultaneously. We describe this phenomenon as the "spike." The above-mentioned algorithms do not take care of this problem. However, this algorithm seeks to overcome the problem with the incorporation of a spike detection mechanism. When a spike is detected, the algorithm switches to 'SPIKE' mode and later reverts back to 'NORMAL' mode when the network condition becomes normal. The basic concept of this algorithm is similar to the EXP-AVG algorithm. However, the value of $\alpha$ is set to 0.875 here.

> In a VoIP system, some packets may suddenly suffer from high end-to-end delay. The result is that no voice packet reaches receiver for some time followed by arrival of a large number of packet reaching almost simultaneously. We describe this phenomenon as "spike."

The Spike-Det algorithm can be summarized by the following steps, where $d_i$ is the delay estimate of the $i$th packet and var is the variation in delay.

1.   $n_i = i^{th}$ packet network delay;

2.   IF (mode == NORMAL) {

   if $(abs(n_i - n_{i-1}) > abs(v_i) * 2 + 800)$

   var = 0; /* Detected beginning of spike */ mode = SPIKE;

   }

   ELSE {

   var = var/2 + $abs((2n_i - n_{i-1} - n_{i-2})/8)$; if (var <= 63) {

   mode = NORMAL; /* End of spike */

   $n_{i-2} = n_{i-1}; n_{i-1} = n_i;$

   return;

   }

   }

3.   IF (mode == NORMAL)

   $d_i = 0.125 * n_i + 0.875 * d_{i-1};$

   ELSE

   $d_i = d_{i-1} + n_i - n_{i-1};$

   $v_i = 0.125 * abs(n_i - d_i) + 0.875 * v_{i-1};$

4.   $n_{i-2} = n_{i-1}; n_{i-1} = n_i;$

   Return;

## 8.3.5   *Window Algorithm [15]*

This algorithm collects the network delays of previously received packets and uses them to estimate the playout delay of the incoming packets. The network delays of last few packets are collected and the delay distribution is updated with every incoming talkspurt. The playout delay of the incoming packet is chosen by obtaining a delay that represents a given percentile among the last few received packets. The determination regarding the playout delay is made with the help of an incrementally updated histogram. When a new packet arrives, its delay is added to the histogram, and the delay of the oldest packet is removed. This algorithm also detects spikes. On detection of a spike, the algorithm stops collecting packet delays. If a talkspurt starts during a spike, then the delay of the first packet of the talkspurt is used as the playout delay for that talkspurt. The efficiency of determination of playout delay for this algorithm depends on the window size, i.e., the number packets considered for recording their delay. If the window is too small, then the estimation of playout delay is likely to be poor. On the other hand, if the window size is too large, large memory is wasted for keeping tracks of long and unnecessary history. The window algorithm is defined by the following steps. Here, $d_i$ is the network delay of the $i$th packet and $p_i$ is the playout delay of the $i$th packet. Two parameters head and tail are used in this algorithm in detecting the beginning and end of a spike, respectively. curr-delay is the given percentile point based on the previous statistics of packet delays.

```
IF (mode == SPIKE)
        IF (dᵢ ≤ tail * old_d) /* the end of a spike */
                mode == NORMAL;
ELSE
        IF (dᵢ > head * pᵢ -1) {      /* the beginning of a spike */
                mode = SPIKE;
                old_d = pᵢ -1;/* save pᵢ to detect the end of a spike  later*/ }
        ELSE {
                IF (delays[curr_pos] ≤ curr_delay)
                        count = count - 1;
                distr_fcn[delays[curr_pos]] = distr_fcn[delays[curr_pos]] - 1;
                delays[curr pos] = dᵢ;
                curr_pos = (curr_pos+1) % w;
```

```
                        distr_fcn[d_i] = distr_fcn[d_i] + 1;
                        IF (delays[curr_pos ] < curr_delay)
                                count = count + 1;
                        WHILE (count < w * q) {
                                curr_delay = curr_delay + unit;
                                count = count + distr_fcn[curr_pos];
                        }
                WHILE (count > w * q) {
                        curr_delay = curr_delay - unit;
                        count = count - distr_fcn[curr_pos];
        }       }
```

## 8.4    The Simulation Setup

We have created the congested network scenario used for the analysis of the above-mentioned adaptive jitter buffer playout algorithms and to assess our new algorithm with the help of OPNET Modeler 14.5.A. The setup consists of four nodes. Two ethernet4_slip8_gtwy_adv gateways are used to interface IP cloud to the communicating nodes. The IP cloud serves the purpose of simulating the presence of an IP backbone in the communication path of the nodes. The gateways and the IP cloud are connected with PPP_adv link whose data rate can be altered.

It is seen from Fig. 8.2 that one of the nodes acts as the Voice caller, whereas another node acts as the Voice callee. These two nodes exchange voice packets between each other. Both these nodes are configured to use G.726 ADPCM coder with 32 kbps bit rate and it produces traffic at a constant rate. The other two nodes, i.e., node 1 and node 2, interchange packets unrelated to the VoIP communication, i.e., the cross traffic. Their communication bit rate varies randomly every second between the lower and upper extremes of 0 and 1000 kbps, respectively. The basic purpose of these nodes is to congest the links between the gateways and the IP cloud. It is worth mentioning that in order to simulate various network behavior, we have simulated the network several times with the capacity of the PPP_adv link having the values of 600, 800, 1000, 1200, and 1400 kbps. These values enable us to study how the VoIP communication behaves when the cross traffic bit rate exceeds the link capacity and then again reduces below the capacity of the link, as in case of 600 and 800 kbps. In case of link capacity of 1000, 1200, and 1400 kbps, the cross traffic is always below the link capacity since the maximum value that can be attained by the bit rates of node 1 and node 2 is 1000 kbps. This is how we have created a varying network, so as to induce variable end-to-end delay to the voice packets exchanged between the pair of voice nodes. The IP cloud serves to simulate

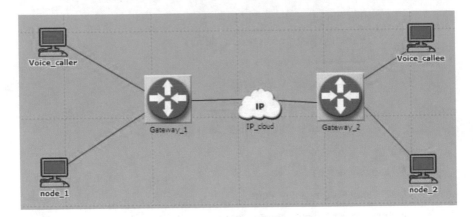

**Fig. 8.2** OPNET simulation setup

the routing functionalities and can also increase the delay and packet loss rate. For simplicity, only the results with network capacities 600, 1000, and 1400 kbps are shown as they cover the three types of jitter conditions, i.e., a network with high jitter (600 kbps network), a network with moderate jitter (1000 kbps network), and a network with low jitter (1400 kbps network).

> We have created a varying network, so as to induce variable end-to-end delay to the voice packets exchanged between the pair of voice nodes.

## 8.5   Analysis of the Existing Adaptive Jitter Buffer Playout Algorithms

The link capacity has been set in accordance with each of the above-mentioned values and VoIP call simulations have been carried out between the two voice nodes to study their behavior. The end-to-end delay values for each of the voice packets were noted. Later, these set of readings have been used to implement the various algorithms and then we have compared the results to find out the improvement in VoIP performance, i.e., reduction in jitter.

It is observed from Table 8.1 that for a network which induces high jitter to the voice packets passing through it (network capacity of 600 kbps), the F-EXP-AVG algorithm discards the least number of packets, and hence the lowest discard ratio.

**Table 8.1** Results for the algorithms for different network capacities

| Network capacity (kbps) | Algorithm | Avg. delay (ms) | Discard ratio (%) | Avg. MOS |
|---|---|---|---|---|
| 600 | EXP-AVG | 493.94 | 6.289 | 1.0232 |
| | F-EXP-AVG | 1446.76 | 0.482 | 1.0041 |
| | Min-D | 417.80 | 8.867 | 1.0191 |
| | Spike-Det | 338.17 | 6.318 | 1.0461 |
| | Window | 341.49 | 5.524 | 1.0320 |
| 1000 | EXP-AVG | 119.83 | 3.928 | 1.9114 |
| | F-EXP-AVG | 276.46 | 0.453 | 2.1900 |
| | Min-D | 112.12 | 6.547 | 1.6462 |
| | Spike-Det | 115.12 | 6.867 | 1.6143 |
| | Window | 103.63 | 3.865 | 1.9386 |
| 1400 | EXP-AVG | 89.52 | 0.756 | 2.8233 |
| | F-EXP-AVG | 102.93 | 0.049 | 3.4264 |
| | Min-D | 88.53 | 2.681 | 2.1664 |
| | Spike-Det | 87.69 | 4.728 | 1.8486 |
| | Window | 86.10 | 0.516 | 2.9840 |

However, it increases the playout delay to such an extent that the average delay increases beyond a tolerable value and hence the voice quality degrades.

The other algorithms induce lower average delay, but discard a large number of voice packets since the packets arrive after the estimated playout time. As a result, the voice call standards go below tolerable limits. The average Mean Opinion Score (MOS) reflects the voice quality offered by each of the algorithms. It is evident that none of the algorithms perform satisfactorily under high jitter conditions.

In a network with moderate congestion (network capacity of 1000 kbps) and consequently moderate jitter, the average delay induced by the algorithms decreases considerably. However, the F-EXP-AVG still imparts higher delay to the voice packets, whereas the Min-D and Spike-Det discard a large number of packets, thereby suffering from large losses. Further, the performances of the algorithms in a network with low congestion (network capacity 1400 kbps) are also tabulated. Here, we can say that since the inter-arrival jitter for the packets is low, the playout algorithms do not incorporate a significant playout delay to the voice packets. Hence, the end-to-end delay does not increase much. However, we can see that the Spike-Det and Min-D algorithms discard a high percentage of packets, and as a result, the call quality provided by them gets degraded.

## 8.6   Proposed Adaptive Jitter Buffer Playout Algorithm

After extensive analysis of some of the existing jitter buffer algorithms, we have come to the conclusion that, when the jitter imparted by the network to the voice packets is very high, the playout delay increases considerably. Moreover, the packet discard ratio also increases beyond tolerable limits. The net result of the above two factors is degradation in the quality of voice in the VoIP session. Our algorithm seeks to reduce the playout delay and packet discard ratio.

> The proposed algorithm seeks to reduce the playout delay and packet discard ratio.

Our algorithm can be summarized in the following steps which are to be followed as long as the VoIP call continues.

**Proposed Algorithm**

1.  $n_i$ = ith packet network delay, $\alpha$ =0.875;

2.  DD = abs($n_i$ - $n_{i-1}$); /* DD indicates the absolute value of the difference in network delay of 2 consecutive packets*/

3.  IF (i < w)          /* w indicates the number of packets to be considered or the window size */

      find out the inter-quartile range of 'i-1' packets;

   ELSE

      find out the inter-quartile range of the last 'w' packets;

4.  IF (inter-quartile range < 5)

      IF ($\alpha$ +0.01< 0.998002)

         $\alpha = \alpha$ +0.01;

   ELSE

         $\alpha = 0.998002$;

ELSE

IF ($\alpha - 0.05 > 0.75$)

$\alpha = \alpha - 0.05$;

ELSE

$\alpha = 0.75$;

5.    IF(mode == NORMAL)

IF ($DD > (1 - \alpha) * 100$)

mode = SPIKE;

ELSE

goto step 6;

ELSE IF ($n_i < \alpha * n_{i-1}$) /* that is, mode = SPIKE */

mode = NORMAL;

Else

goto step 6;

6.    IF(mode ==SPIKE)

$d_i = 0.75 \times d_{i-1} + (1-0.75) \times n_i$;

ELSE /* that is, mode = NORMAL */

IF(new talkspurt)

$d_i = n_i$;

ELSE

$d_i = \alpha * d_{i-1} + (1 - \alpha) \times n_i$;

7.    $Y = abs(d_i - n_i)$ ;

8.    $v_i = 0.998002 \times v_{i-1} + (1-0.998002) \times Y$;

9.    Playout delay = $d_i + 4 v_i$;

The algorithm is pictorially represented in Fig. 8.3. We estimate the network characteristics by keeping track of the last 'w' received packet. 'w' represents a sliding window. This sliding window helps the receiver to estimate the playout delay of the future packets from the end-to-end delays of the previous 'w' received packets. Choosing a small value for 'w' may lead to improper estimation of playout delay, whereas a high value of 'w' may lead to wastage of memory by keeping track of unnecessary packets. Hence, the value of 'w' should be chosen carefully to get the best results. We accordingly vary the value of $\alpha$, where $\alpha$ is a parameter that determines how much a newly received packet depends on the previously received packets.

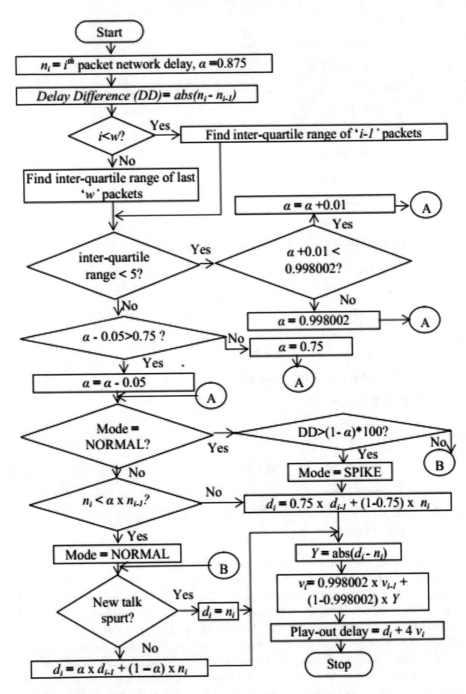

**Fig. 8.3** Flowchart of the proposed adaptive jitter buffer playout algorithm

> The sliding window helps the receiver to estimate the playout delay of the future packets from the end-to-end delays of the previous 'w' received packets.

## 8.7 Results

The QoS of a VoIP call can be best described by the MOS value as both the end-to-end delays of the packets and the packet loss are considered for the calculation of the MOS. The algorithm was initially implemented by setting the value of 'w' as 100. Following this development, the effect of varying window sizes on the performance of the algorithm was measured. Finally, a comparative analysis of the performance of the proposed algorithm with the performances of the analyzed algorithms has been made in order to assess the improvement offered by the implementation of the proposed adaptive jitter buffer playout algorithm. The results obtained have been included in the following subsections.

### 8.7.1 Implementing the Algorithm with a Window Size of 100

Initially, the algorithm used the data from the latest 100 packets in order to estimate the recent trends in the network characteristics. Several simulations were being carried out to find out the effectiveness of the proposed algorithm and it gave a better MOS than the other discussed algorithms. The results also show that it reduced jitter considerably.

It can be seen from Table 8.2 that the proposed algorithm performs satisfactorily for all of the above network scenarios. However, the MOS is low for a network bandwidth of 600 kbps because the network congestion is very high in this case. As a result, there is a very large jitter between the consecutive voice packets. So, in order to compensate for jitter, the algorithm incorporates a significant playout delay, as a

**Table 8.2** Results for the proposed algorithm for different network capacity with a window size of 100

| Bandwidth (kbps) | Avg. delay (ms) | Packet discard ratio (%) | Average MOS | Percentage reduction in jitter |
|---|---|---|---|---|
| 600 | 276.93 | 2.635 | 1.4508 | 35.34 |
| 800 | 161.42 | 2.846 | 2.0657 | 38.96 |
| 1000 | 107.80 | 2.806 | 2.1059 | 44.86 |
| 1200 | 92.50 | 0.664 | 2.8709 | 50.19 |
| 1400 | 87.68 | 0.171 | 3.2932 | 54.49 |

result of which, end-to-end delay is increased. Moreover, due to the high unpredictable nature of the network, some packets arrive later than it is estimated and
hence they are rejected. So, the discard ratio is also high. Both these factors degrade
voice quality, and hence, the average MOS rating is also reduced. However, as the
capacity of the network increases, the probability of the network becoming congested is also lowered. As a result, the jitter also decreases with increase in network
capacity. This enhances VoIP performance. The algorithm, however, still incorporates a small playout delay, in order to reduce the jitter further.

In Fig. 8.4, it can be seen that the jitter reduces significantly following the
application of the proposed adaptive jitter buffer playout algorithm. The average
jitter throughout the duration of the call falls from 18.38 to 11.88 ms. The above
result reflects the effectiveness of the algorithm under congested network that
imparts high jitter to the voice packets passing through it. Figures 8.5 and 8.6

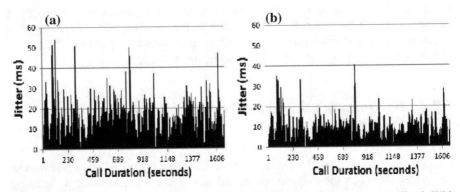

**Fig. 8.4**  Inter-arrival jitter for network capacity of 600 kbps. **a** Without playout buffer. **b** With
proposed algorithm (window size 100)

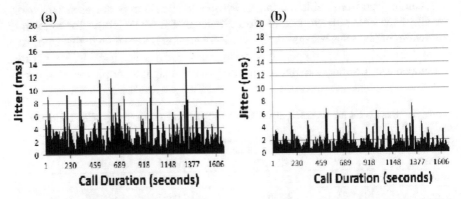

**Fig. 8.5**  Inter-arrival jitter for network capacity of 1000 kbps. **a** Without playout buffer. **b** With
proposed algorithm (window size 100)

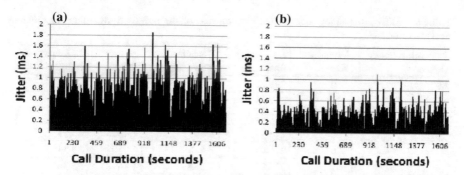

**Fig. 8.6** Inter-arrival jitter for network capacity of 1400 kbps. **a** Without playout buffer. **b** With proposed algorithm (window size 100)

further illustrate the behavior of the algorithm to moderate and low jitter, where the network capacity is 1000 and 1400 kbps, respectively. It is seen from the figures that the proposed algorithm reduces jitter considerably and performs well in all three network scenarios.

The proposed algorithm reduces jitter considerably and performs well in all three network scenarios.

## 8.8 Effect of Various Window Sizes

After establishing the effectiveness of our proposed adaptive jitter buffer algorithm, we conducted further experimentations in order to find out the optimum window size for all network conditions. For this purpose, we have conducted exhaustive simulations using various window sizes. We have primarily focused on maximizing the MOS of the VoIP call as it reflects the call quality from the user's point of view. However, we have also considered the other important parameters that reflect the QoS of a call from the technical point of view. These parameters include average delay, packet discard ratio, and percentage improvement in jitter. The effects of window size on the VoIP call performance have been illustrated in Table 8.3.

We have primarily focused on maximizing the MOS of the VoIP call as it reflects the call quality from the user's point of view.

From Table 8.3, we can clearly see that for a network capacity of 600 kbps, the best improvement in jitter is obtained for a window size of 40. However, the best average MOS value is obtained with a window size of 60. As the network capacity

**Table 8.3** Effect of window size on the performance of the proposed algorithm for different network capacities

| Network capacity (kbps) | Window size | Avg. delay (ms) | Discard ratio (%) | Avg. MOS | % Reduction in jitter |
|---|---|---|---|---|---|
| 600 | 0 | 291.71 | 2.238 | 1.4405 | 29.27 |
|  | 20 | 277.13 | 2.635 | 1.4447 | 35.51 |
|  | 40 | 277.02 | 2.635 | 1.4451 | 35.37 |
|  | 60 | 276.97 | 2.635 | 1.4508 | 35.35 |
|  | 80 | 276.93 | 2.635 | 1.4508 | 35.34 |
|  | 100 | 276.93 | 2.635 | 1.4508 | 35.34 |
| 800 | 0 | 163.55 | 3.201 | 1.9922 | 33.56 |
|  | 20 | 161.27 | 2.896 | 2.0571 | 39.26 |
|  | 40 | 161.83 | 2.846 | 2.0540 | 38.84 |
|  | 60 | 160.98 | 2.820 | 2.0712 | 38.86 |
|  | 80 | 161.47 | 2.846 | 2.0656 | 38.93 |
|  | 100 | 161.42 | 2.846 | 2.0657 | 38.96 |
|  | 0 | 107.96 | 2.701 | 2.1333 | 42.17 |
|  | 20 | 108.14 | 2.860 | 2.1012 | 51.10 |
|  | 40 | 107.93 | 2.592 | 2.1561 | 48.80 |
|  | 60 | 108.04 | 2.699 | 2.1335 | 46.85 |
|  | 80 | 108.28 | 2.672 | 2.1387 | 45.74 |
|  | 100 | 107.80 | 2.806 | 2.1059 | 44.86 |

increases, the optimum window size seems to decrease. For a network with capacity of 800 and 1000 kbps, the highest improvement in jitter is obtained for a window size of only 20. But, the best MOS is observed for a window size of 40 and 60, respectively. When looking at the least average delay and packet discard ratio for various window sizes, we can clearly see the algorithm gives the best results for the window sizes between 40 and 60.

It may be noted that the results for network capacities of 1200 and 1400 kbps have not been included in Table 8.3 because the values remain constant even though there is change in the window size. This is primarily due to the fact that these two scenarios induce very low jitter to the passing voice packets. Hence, the playout delay mostly depends on the end-to-end delay of the received packet only. So, the window size does not have any effect on the performance of the algorithm.

The results can be further verified by Figs. 8.7a, b and 8.8a, b. Figure 8.7a reflects the effect of window size on average delay of the voice packets for the duration of the voice call. It is seen that the delay is high if a zero window size is used. However, as the window size is increased to 10, the delay decreases considerably and remains fairly constant with further increase in window size. Figure 8.7b on the other hand shows that the window size has a profound effect on the packet discard ratio. It is clearly seen that the best results in terms of packet discard ratio are obtained for a window size of 40. Figure 8.8a provides illustration for the most important of the VoIP QoS parameters, the MOS. It is seen that the

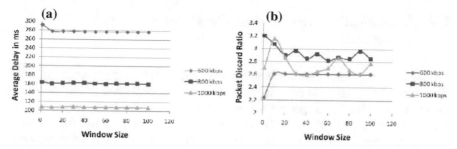

**Fig. 8.7** A comparison of various parameters w.r.t. window size. **a** Average end-to-end delay and **b** packet discard ratio

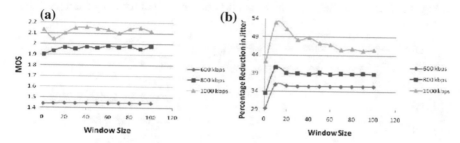

**Fig. 8.8** A comparison of various parameters w.r.t. window size. **a** MOS. **b** Percentage reduction in jitter

MOS depends on the size of the window used for the operation of the proposed adaptive jitter buffer algorithm. It is seen that the best MOS values are obtained for around two window sizes, i.e., 40 and 80. Since choosing a lower window size helps to reduce the memory consumption and the processing delay, it is better to choose a window size of 40 as an optimum value.

Figure 8.8b provides the analysis results of the dependency of percentage improvement of jitter with respect to changing window sizes. Here, we find that the best results are obtained for a window size of 20. However, the best MOS value is not obtained for a window size of 20, and hence, the algorithm will fail to provide its full potential if we resort to a window size of 20. Hence, we sacrifice a little improvement in terms of jitter in order to provide a more satisfactory user experience. Thus, after a thorough analysis we come to the conclusion that a window size of 40 can be considered as the optimum value for a seamless operation of the proposed algorithm.

> Since choosing a lower window size helps to reduce the memory consumption and the processing delay, it is better to choose a window size of 40 as an optimum value.

## 8.9    Comparative Results with the Other Analyzed Algorithms

Further endeavors have been taken in order to find out where the proposed adaptive jitter algorithm stands, when compared with the performances of the other discussed algorithms. Our algorithm has given the lowest end-to-end delay among all the algorithms especially when the network is more congested and the extent of jitter in the voice packets is very high. The end-to-end delay results can be observed in Fig. 8.9. When examined for packet discard ratio, it has been found that our algorithm performs quite well. However, the F-EXP-AVG algorithm has even lower packet discard ratio. The packet discard ratio comparison has been illustrated in Fig. 8.10. Upon further examinations, it is observed that for congested mediums,

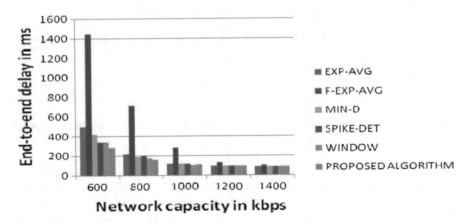

**Fig. 8.9**  Comparison of the end-to-end delays of the different algorithms

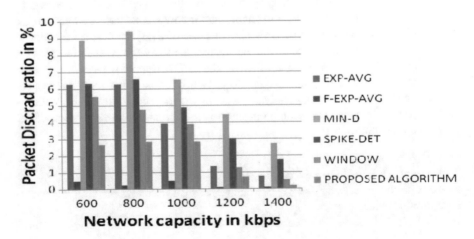

**Fig. 8.10**  Comparison of the packet discard ratio of the different algorithms

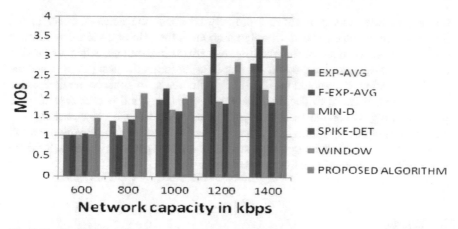

**Fig. 8.11** Comparison of MOS of the different algorithms

our algorithm gives the best MOS values. However, for network with very low jitter, F-EXP-AVG gets the edge because of its lower packet discard ratio. In a nutshell, we can say that the existing algorithms aim to minimize either end-to-end delay or packet loss. While one of the above-mentioned factors is reduced, the other gets worse. As a result, voice quality gets degraded. Our algorithm aims to find a trade-off between these two factors. Hence, it provides the best voice quality. The comparative results of MOS values have been included in Fig. 8.11.

> In a nutshell, the existing algorithms aim to minimize either end-to-end delay or packet loss. While one of the above-mentioned factors is reduced, the other gets worse. As a result, voice quality gets degraded. Our algorithm aims to find a trade-off between these two factors. Hence, it provides the best voice quality.

## 8.10 Summary

A congested network transports voice packets with uneven delay. The result of this unevenness is incorporation of jitter in the consecutive voice packets. Jitter is not desirable during a voice call as it leaves the user dissatisfied. Several, algorithms have already been proposed to add a further playout delay to the voice packets in hope of minimizing the jitter. However, selecting the optimum playout delay is a tricky part. These algorithms often under-estimate or over-estimate the network delay of future incoming voice packets, resulting in discarding of the packets or long undesirable end-to-end delay, respectively. We have proposed an algorithm that addresses to this problem and properly estimates the network delay of the future incoming voice packets. Our algorithm takes the help of a sliding window in order to get an estimate of the variation in network conditions. After conducting extensive analysis, we have found out that in order to get the best performance the optimum window size should be 40. Our algorithm aims to enhance the QoS of the VoIP session. It seeks to decrease the end-to-end delay and packet discard ratio while allowing a tolerable amount of jitter to be present in the voice packets. The primary aim of our algorithm is to enhance user experience by improving the MOS of the call. We are conducting further studies in order to get even better QoS for the voice calls in a congested network scenario.

# References

1. J. Davidson, J. Peters, *A Systematic Approach to Understanding the Basics of VoIP, Voice over IP Fundamentals* (CISCO press, Indianapolis, 2000)
2. J.C. Bolot, A. Vega-Garcia, Control mechanisms for packet audio in the Internet, in *Proceedings of IEEE Annual Joint Conference of the IEEE Computer Societies. Networking the Next Generation*, pp. 232–239, San Francisco, USA, 24–28 Mar 1996
3. K. Wallace, *Implementing Cisco Unified Communications Voice over IP and QoS (Cvoice) Foundation Learning Guide* (Cisco Press, Indianapolis, 2011)
4. C. Sanghyun, B.F. Womack, QoS-based adaptive playout scheduling based on the packet arrival statistics: Capturing local channel characteristics, in *Proceedings of 2010 IEEE International Workshop Technical Committee on Communications Quality and Reliability*, June 2010. https://doi.org/10.1109/cqr.2010.5619942
5. A. Mukhopadhyay, T. Chakraborty, S. Bhunia, I. Saha Misra, S.K. Sanyal, Study of enhanced VoIP performance under congested wireless network scenarios," in *Proceedings of Third International Conference on Communication Systems and Networks* (COMSNETS 2011), pp. 1–7, Bangalore, India, 4–8 Jan 2011. https://doi.org/10.1109/comsnets.2011.5716509
6. M.F. Hong, H.W. Hsu, W.L. Du, Y. H. Hung, M.H. Lee, A fast-startup TCP mechanism for VoIP services in long-distance networks, in *Proceedings of International Conference on Intelligent Information Hiding and Multimedia Signal Processing*, pp 185–188, Dec 2006. https://doi.org/10.1109/iih-msp.2006.264976
7. J. Yan, M. May, B. Plattner, Distributed and optimal congestion control for application-layer multicast: a synchronous dual algorithm, in *Proceedings of 5th IEEE Consumer Communications and Networking Conference*, pp. 279–283, 10–12 Jan 2008. https://doi.org/10.1109/ccnc08.2007.69
8. A.B. Rus, M. Barabas, G. Boanea, Z. Kiss, Z. Polgar, V. Dobrota, Cross-layer QoS and its application in congestion control, in *Proceedings of 2010 17th IEEE Workshop on Local and Metropolitan Area Networks* (LANMAN), pp. 1–6, May 2010. https://doi.org/10.1109/lanman.2010.5507149
9. T. Chakraborty, A. Mukhopadhyay, I. Saha Misra, S.K. Sanyal, Optimization technique for configuring IEEE 802.11b access point parameters to improve VoIP performance, in *Proceedings of International Conference of Computer and Information Technology*, Dhaka, Bangladesh, 23–25 Dec 2010. https://doi.org/10.1109/iccitechn.2010.5723919
10. P. Gournay, K.D. Anderson, Performance analysis of a decoder-based time scaling algorithm for variable jitter buffering of speech over packet networks, in *Proceedings of IEEE International Conference on Acoustics, Speech and Signal Processing*, May 2006. https://doi.org/10.1109/icassp.2006.1659946

11. K.M. McNeill, M. Liu, J.J. Rodriguez, An adaptive jitter buffer play-out scheme to improve VoIP Quality in wireless networks, in *Proceedings of IEEE Military Conference*, Oct 2006. https://doi.org/10.1109/milcom.2006.302119

12. H. Schulzrinne, S. Casner, R. Frederick, V. Jacobson, *RTP: A Transport Protocol for Real-Time Applications*, RFC 3550, July 2003

13. R.G. Cole, J.H. Rosenbluth, Voice over IP performance monitoring. ACM SIGCOMM Comput. Commun. Rev. **31**(2), (2001)

14. R. Ramjee, J. Kurose, D. Towsley, H. Schulzrinne, Adaptive playout mechanism for packetised audio application in wide-area networks, in *Proceedings of INFOCOM '94. Networking for Global Communications*, pp. 680–688 vol. 2, 12–16 Jun 1994 https://doi.org/10.1109/infcom.1994.337672

15. S.B. Moon, J. Kurose, D. Towsley, Packet audio playout delay adjustment: performance bounds and algorithms. Acm/Spring Multimedia Syst. **6**(1), 17–28 (1998). https://doi.org/10.1007/s005300050073

16. J. Postel (ed.), *Transmission Control Protocol specification: ARPANET Working Group Requests for Comment*, RFC 793 (1981)

17. V. Jacobson, Congestion avoidance and control, in *Proceedings of 1988 ACM SIGCOMM Conf.*, pp. 314–329, Stanford (1988)

# Chapter 9
# Adaptive Packetization Algorithm to Support VoIP Over Congested WLANS

## 9.1 Introduction

Effective voice communication techniques hold a key position for successful implementation of VoIP as a service. However, the network over which VoIP operates is time variant. It may become congested if a linked is choked up with traffics exceeding the traffic handling capacity of the underlying network. Inspired by the works of Ngamwongwattana [1], we aim to exploit the use of variable voice packet payload sizes for implementing an adaptive VoIP strategy. The result is achieved by varying the number of voice sample frames in the payload of the RTP packets based on the network condition. For this purpose, we have created a snapshot of a network with the help of OPNET Modeler 14.5.A and performed exhaustive simulations in order to analyze the effect of congestion on an ongoing VoIP call.

> We aim to exploit the use of variable voice packet payload sizes for implementing an adaptive VoIP strategy.

Moreover, we have also studied the effects of using variable packet sizes in different network congestion scenarios. The inferences drawn from the above simulations have helped us to propose an adaptive packetization algorithm that alters the number of voice sample frames in the RTP packet payload depending upon the network congestion. The algorithm indicates well-defined steps following which minimizes the end-to-end delay and enhance Mean Opinion Score (MOS). Further, we have implemented the adaptive algorithm in a voice node with the help of OPNET Modeler 14.5.A. Extensive simulations have been performed in order to assess the performance of the proposed algorithm under varying network congestion scenarios.

© Springer International Publishing AG, part of Springer Nature 2019
T. Chakraborty et al., *VoIP Technology: Applications and Challenges*,
Springer Series in Wireless Technology, https://doi.org/10.1007/978-3-319-95594-0_9

## 9.2   Background Literature

VoIP is a relatively new technology that makes use of the already established Internet backbone to transport the voice packets. The primary protocol that drives the operations of the Internet in the network layer is the Internet protocol (IP). IP is responsible for moving the data packets through an interconnected set of networks [2]. It is an unreliable, connectionless protocol providing best-effort service. IP uses error detection mechanisms and discards corrupted packets. IP does its best to deliver a data packet to its proper destination; however, it does not give any guarantee that the packet will be delivered correctly [3]. Moreover, it treats each of the data packets individually and routes them accordingly without taking into account the other packets. This implies that each packet may follow different route from the source to the destination and may arrive out of order [3].

In the transport layer, the Internet makes use of either of the two protocols for its functioning; the TCP or the User Datagram Protocol (UDP). TCP is a stream connection-oriented reliable transport protocol. TCP provides a reliable orderly delivery of a stream of bytes from one network node to another. Major applications such as the World Wide Web, email, and file transfer rely upon the TCP for their operations. TCP provides flow and error control. Hence, it is a 'slow' protocol and incorporates significant latency. Moreover, the error control mechanism of TCP makes it inappropriate for VoIP. In VoIP, there is no point in retransmitting a lost or corrupted packet since it upsets the procedure of sequencing and time stamping of the voice packets [3]. Moreover, retransmission of packets further swamps an already congested network. As a result, VoIP relies on UDP to carry out its transport layer operations. UDP is more suitable for VoIP as it supports multi-casting and has no retransmission strategy. However, UDP has no provision of time stamping or sequencing of the voice packets. In order to get around this problem, VoIP utilizes two more protocols working in tandem; the RTP and the Real-time Transmission Control Protocol (RTCP).

RTP provides sequence numbers and time stamps to the voice packets. These help the receiver to reorder the voice packets, as the voice packets may follow different paths from the source to the destination and arrive out of order due to the packet-wise approach of the IP. RTCP plays a very important role in congestion control. RTCP is based on the periodic transmission of control packets to all participants in the session, using the same distribution mechanism as that of the data packets [4]. Primarily, RTCP provides the feedback on the quality of data distribution. This feedback information obtained from the RTCP receiver reports (RR). The RRs contain information about the fraction lost, cumulative packet loss, inter-arrival jitter, etc. This feedback information allows the source to estimate the loss rate experienced at the receiver end and provides a guideline for adjusting its transmission rate. Moreover, RTCP allows the assignment of a unique Canonical Name (CNAME) to each participant in a VoIP session. The CNAME acts as a unique identifier and helps the individual participants to keep track of the number of VoIP users. However, if the number of participants in the session is very large, then the number of RTCP packets exchanged between the participants should be controlled and must not be allowed to exceed a certain value because it may lead to an even worse congested network.

Packetizer is a type of mechanism, which follows the encoder. Its function is to place voice frames after encoding into RTP/RTCP-, UDP-, and IP-contained packets. De-packetizer reverses this process before VoIP packets are decoded. Their functionalities in a VoIP System are clearly illustrated in Fig. 9.1.

The processes of digitizing and packetizing the received analog voice sample to make it compatible with the Internet require time. This processing delay is significant enough to be taken into account in the end-to-end delay calculation. Moreover, the received packet must also be de-packetized and then converted into analog form so that the intended receiver comprehends the information. These processing delays along with the time required by the voice packet to traverse the network in order to reach the receiver from the sender comprise of the end-to-end delay. The end-to-end delay must be kept within 150 ms so that the VoIP user is comfortable with the quality of the voice call [5].

As of now, alterations in the voice packet payload size have already been explored in order to enhance VoIP performance. The inspiration behind using packetization as a means for adaptive VoIP strategy can be found in the packet format of a VoIP packet. With a closer look into the packet format, we can see that at least 40 bytes of overhead is incurred in order to transmit a VoIP packet. To be more elaborate, 20 bytes of IP header, 8 bytes of UDP header, and 12 bytes of RTP header comprise the overhead. For a G.726 codec with a bit rate of 32 kbps, the codec sample size is 20 bytes and the VoIP payload size is 80 bytes [1, 6]. So, it is clear that a large percentage of the transmitted information comprises the header overhead. Using [1, 2, 6], the encapsulation of VoIP packets can be illustrated as in Fig. 9.2.

The idea can be further illustrated by the following equations. Let the number of voice sample frames in the RTP packet be $n$. We know that the sample frame size for G.726 codec with a bit rate of 32 kbps is 20 bytes.

$$\text{So, the size of the RTP packet} = 20n + 12 \text{ bytes} \qquad (9.1)$$

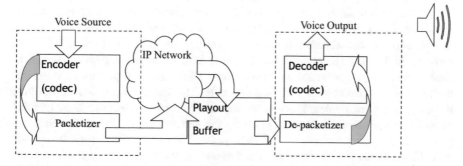

**Fig. 9.1** Packetizer/de-packetizer in a VoIP system

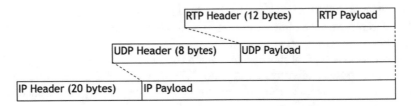

**Fig. 9.2** Encapsulation procedure of voice packets

$$\text{The size of the UDP datagram} = (20n + 12) + 8\,\text{bytes} \qquad (9.2)$$

$$\text{The size of the IP packet} = (20n + 20) + 20\,\text{bytes} \qquad (9.3)$$

$$\text{Therefore, the payload-to-overhead ratio} = 20n/40 = n/2 \qquad (9.4)$$

Equation (9.4) indicates that as we increase the number of voice sample frames in the RTP packet the payload-to-overhead ratio increases. As a result, we can send more data by using lower number of bits. This reduces network congestion.

> As we increase the number of voice sample frames in the RTP packet, the payload-to-overhead ratio increases. As a result, we can send more data by using lower number of bits. This reduces network congestion.

## 9.3  The Simulator

Optimized Network Engineering Tool (OPNET) [7] provides comprehensive development environment supporting the modeling of communication networks and distributed systems. Both behavior and performance of the modeled systems can be analyzed by discrete event simulation as described in Chap. 4. Tool for all the phases of our study including model design, simulation, data collection, and data analysis is incorporated in OPNET environment. Various constructs pertaining to communication and information processing are provided by OPNET. Thus, it provides high leverage for modeling and distributed systems. Graphical specifications of a model are provided by OPNET most of the times. It provides a graphical editor to enter the network and model details. These editors provide an intuitive mapping from the modeled system to the OPNET model specification. OPNET provides four such types of editors namely the network editor, the node editor, the process editor, and parameterized editor organized in a hierarchical way. It supports model level reuse, i.e., models developed at one layer can be used by another model at a higher layer. All OPNET simulations automatically include support for analysis

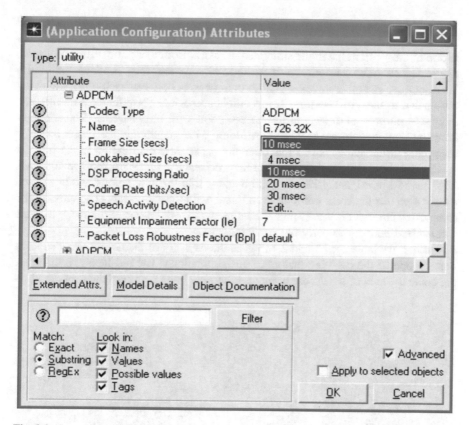

**Fig. 9.3** Screenshot of the voice node's attribute in OPNET

by a sophisticated interactive debugger. Technology developers leverage advanced simulation capabilities and rich protocol model suites to design and optimize proprietary wireless protocols.

In this chapter, we take the advantages of OPNET Wireless modeler suites (OPNET 14.5.A). We have performed a series of simulation with the help of the WLAN model, which provides very reliable results for wireless communication. Figure 9.3 represents the screenshot of the voice node's attributes table.

## 9.3.1   Simulation Setup

We have created a LAN scenario with the help of OPNET Modeler 14.5.A. The setup consists of eight communicating nodes. An access point provides connectivity and routing functions. The traffic handling capacity of the access point can be altered and this very property has been used to create varying network scenarios.

It can be seen from Fig. 9.4 that one of the nodes acts as the VoIP caller and another node acts as the VoIP callee. These two nodes are the source and destination of the VoIP traffic in the simulated network. Both these nodes are configured to use G.726 ADPCM coder with 32 kbps bit rate and they produce traffic at a constant rate. The other nodes exchange data packets between each other which are not related to the VoIP session. These unrelated packets form the cross traffic and their sole purpose is to swamp the access point with heavy traffic. These nodes, however, can initiate a VoIP session along with generating the cross traffic if required. Earlier studies have shown that the packet size distribution in the Internet is centered about three values. To be more precise, 60% of the packets are 40 bytes, 25% are 550 bytes, and 15% are 1500 bytes [8]. In the simulations, we have used packet sizes for the cross traffic in accordance with these findings. The cross traffic also passes through the access point, creating a network bottleneck. Thus, we create a congested network to carry out our simulations.

Further, we have used 1 and 2 Mbps as the capacity of the access point so as to make it one of the deciding factors in our simulations. The red hexagon indicates the coverage area of the access point and the blue lines represent the cross load on

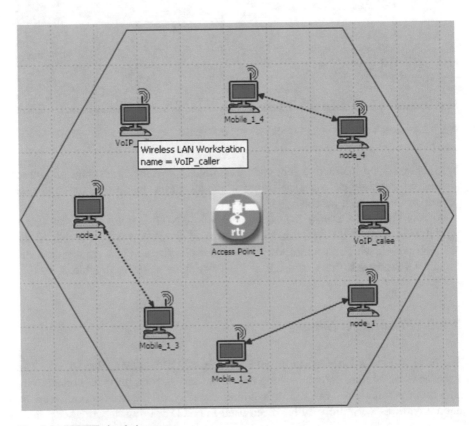

**Fig. 9.4** OPNET simulation setup

the wireless access point. We have varied the cross traffic as a percentage of the access point's traffic handling capacity and studied its impact on the VoIP session. Next, we have altered the number of voice samples in the RTP packet from 1 to 10 and measured its impact on the VoIP session for each of the above-mentioned cross traffic load. Finally, we have increased the number of participants of the VoIP session. Thus, we have investigated how the network behaves for different levels of network congestion, packetization and number of participants in a VoIP session. In order to get more results in similar conditions but for a network with higher capacity, we have changed the traffic handling capacity of the access point from 1 to 2 Mbps.

## 9.3.2  Simulation Scenarios

The simulation setup is being set so as to represent a congested network scenario. In order to analyze the VoIP call quality under different network conditions, we have carried out extensive simulations for each of the aforementioned scenarios. We have recorded the average end-to-end delay and the average (time-average) MOS values for each of the simulations. We have concentrated on the end-to-end delay and MOS values because they portray the call quality from the end user's point of view. An end-to-end delay value, above 150 ms, is not acceptable. Similarly, the MOS of a call should be high enough for a pleasing user experience.

At first, we use only two participants for the voice session. For this session, both the participants use G.726 codec with 32 kbps bit rate. The packet payload size is not altered and only the default G.726 packets are used. We increase the cross traffic from 0 to 90% of the link capacity and observe the results.

Figures 9.5 and 9.6 represent the end-to-end delay and MOS of the voice calls with varying cross traffic. It can be seen from Fig. 9.5 that as the cross traffic percentage increases, the end-to-end delay of the voice packets also increases. This

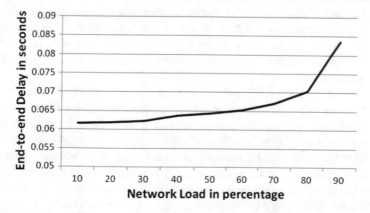

**Fig. 9.5**  Increase in end-to-end delay w.r.t increase in network cross traffic

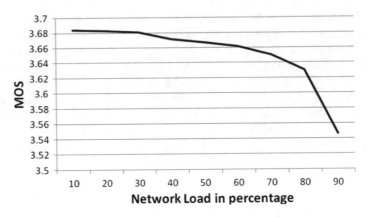

**Fig. 9.6** Decrease in MOS w.r.t increase in network cross traffic

happens because, with the increase in the network cross traffic, the network congestion increases. In our case, the load on the access point increases as it has to handle more and more traffic, or in other words it has to handle more packets. The access point normally processes the packets before retransmitting them. This processing is done in order to route the packets to their correct destinations. Hence, when the packet arrival rate exceeds the packet processing rate of the access point, the access point can no longer retransmit the packet as soon as receiving it. The packets get queued in the access point buffer. This phenomenon incurs significant amount of delay to the passing voice packets. Thus, the end-to-end delay increases with the increase in cross traffic percentage. This effect becomes even more perceptible in a larger network since there are a large number of routers en route from the source to the receiver. Each of the routers has to process the passing packets in order to provide proper routing functionalities. All these processing delays add up to generate a large end-to-end delay.

Figure 9.6 represents the change in MOS with respect to the change in network cross traffic. It is evident from Sect. 9.2 that the MOS depends on the end-to-end delay. The more the end-to-end delay incurred increases, the less the MOS becomes. Since the end-to-end delay increases with the increase in percentage cross traffic, the MOS degrades. Since MOS is a direct correlation with the user experience, it is evident that more and more users are dissatisfied as the cross traffic percentage increases.

It is evident from Figs. 9.5 and 9.6 that for a normal G.726 VoIP call, the call quality degrades with increase in network congestion, and hence, there is a need for a procedure that can provide acceptable call quality even in adverse network conditions. So, we analyze the effect of using altered packet payload size on the quality of a VoIP call. As we increase the number of voice sample frames in the RTP packet payload, we can see that the end-to-end delay decreases significantly as depicted in Fig. 9.7. This happens because with the increase in the number of voice sample frames in the voice packet payload, the payload-to-overhead ratio increases as already shown in Sect. 9.2. In other words, more information gets transmitted

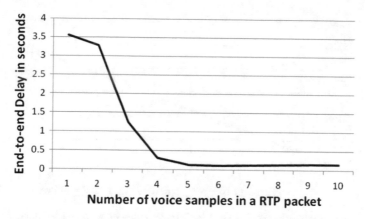

**Fig. 9.7** Effect of packetization on end-to-end delay in a congested network

from the source to the destination by consuming lesser number of bits. This reduces the network congestion significantly.

> As we increase the number of voice sample frames in the RTP packet payload, we can see that the end-to-end delay decreases significantly.

Moreover, sending more sample frames in a single packet signifies that the whole message can be sent by using lesser number of voice packets. As a result, the routers have to deal with lesser number of packets. This means that the processing delay in the intermediate routers gets reduced, and as a result, the end-to-end delay also gets lowered. Using smaller packet payload sizes on the other hand, it generates high end-to-end delay in a low capacity network because the network spends a large time to queue and process the voice packets.

> Using smaller packet payload sizes on the other hand, it generates high end-to-end delay in a low capacity network because the network spends a large time to queue and process the voice packets.

This method, however, has a drawback. As we incorporate more sample frames to a single VoIP packet, we have to wait longer for more sample frames to be generated, and thus, the packetization delay increases as shown in Fig. 9.8. For delay-sensitive VoIP, this is not desirable. In a congested network, using larger packet payloads decreases the end-to-end delay up to a certain optimum packet size. This optimum value primarily depends on the network capacity. After reaching the optimum voice packet size, the network delay does not decrease any further. However, with the incorporation of each voice frame into the voice packet payload,

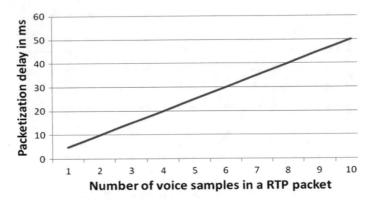

**Fig. 9.8** Effect of increasing number of voice sample frames on packetization delay

the packetization delay increases. Hence, the end-to-end delay also increases, and as a result, the call quality degrades. The results can be further verified in Fig. 9.9 where the MOS variation with increase in voice packet payload size has been charted. It can be seen from Fig. 9.9 that as the number of voice sample frames in a voice packet payload approach the optimum value, the MOS increases considerably. However, on incorporating more voice sample frames in the RTP payload, the MOS degrades. This symbolizes degradation in the call quality. Hence, it is mandatory to choose a trade-off between the network delay and packetization delay in order to get the best call quality.

> Hence, it is mandatory to choose a trade-off between the network delay and packetization delay in order to get the best call quality.

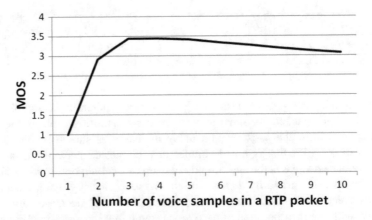

**Fig. 9.9** Effect of packetization on MOS in a congested network

However, it is worth mentioning that it is not prudent to select a specific number of voice sample frames in the RTP packet payload and stick to it for the whole VoIP session. The optimum number of voice samples in the voice packet payload may vary depending on the network condition. As a result, fixing the number of voice sample frames to a definite value will not yield the best results throughout the VoIP session as the network conditions may vary with time. It can be easily concluded from Figs. 9.10a, 9.11a, 9.12a, and 9.13a that we cannot choose the optimum value of packet payload size beforehand. We can see in Fig. 9.10a that for a VoIP session between two users and 80% cross traffic, the optimum number of voice sample frames in a voice packet payload is 1. However, as the numbers of VoIP users increase, i.e., the congestion in the network increases further, the optimum number for the voice sample frames also tends to increase. From Figs. 9.11a, 9.12a, and 9.13a, we can observe that the optimum values for the voice sample frames with 80% cross traffic are 2, 3, and 6 respectively. The result can be further verified from Figs. 9.10b, 9.11b, Fig. 9.12b, and Fig. 9.13b, where the MOS for the VoIP session

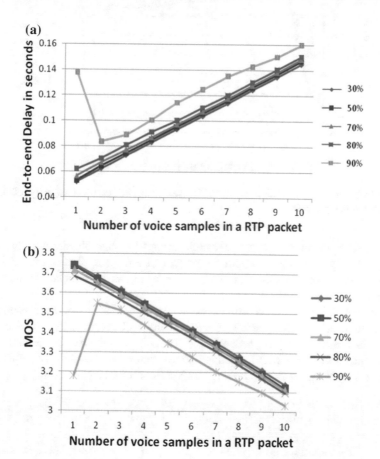

**Fig. 9.10** Effect of packetization on 2 VoIP users with access point capacity of 1 Mbps. **a** End-to-end delay. **b** MOS

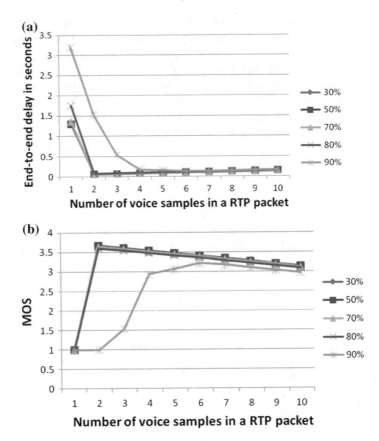

**Fig. 9.11** Effect of packetization on 3 VoIP users with access point capacity of 1 Mbps. **a** End-to-end delay **b** MOS

with respect to the number of voice sample frames has been illustrated. The figures imply that only by using the optimum number of voice sample frames for a given network congestion scenario, the best performance can be obtained from a VoIP system.

However, it can be noted from Figs. 9.12a, b, 9.13a, b that even though the packetization scheme successfully decreases the end-to-end delay for a network with 90% cross traffic, the network condition is so adverse for feasible VoIP communication that the MOS remains unacceptable. For this kind of network situation, it is just impossible to hold a VoIP session of acceptable quality.

> The optimum number of voice samples in the voice packet payload may vary depending on the network condition.

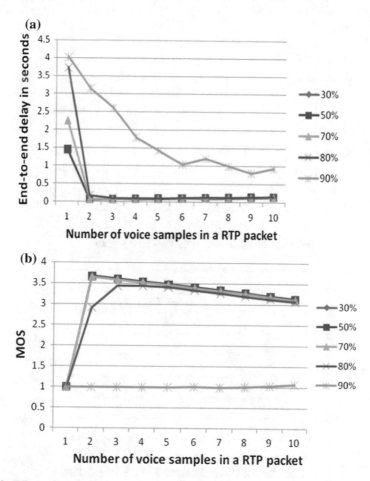

**Fig. 9.12** Effect of packetization on 4 VoIP users with access point capacity of 1 Mbps. **a** End-to-end delay **b** MOS

Following the same procedure for a WLAN access point of traffic handling capacity of 2 Mbps, we get another set of data as shown in Figs. 9.14, 9.15, 9.16, and 9.17. Here, we can observe that the optimum packet payload size has changed even though the other parameters, i.e., number of VoIP users and the percentage of cross traffic, have been the same. Comparing Fig. 9.14a with Fig. 9.10a, we can infer that due to increase in network capacity, the VoIP performance has been enhanced even in networks with high congestion. Figure 9.14b also indicates improvement in VoIP performance in comparison to Fig. 9.10b.

It is apparent from Fig. 9.14a that packetization does not provide any improvement over the VoIP performance. However, here also packetization helps in improvement if we use smaller packet sizes. But, the sample size of the codec used puts a limit to the smallest possible packet payload size we can use. The smallest packet size that can be used by a G.726 codec operating at 32 kbps is 20 bytes. So,

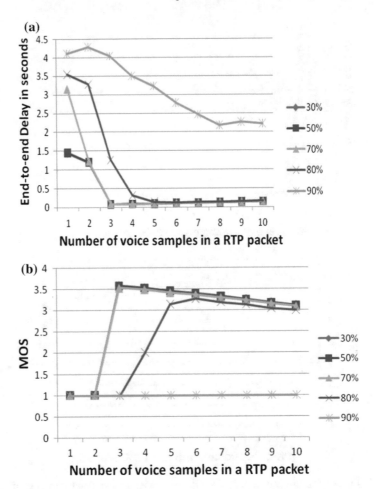

**Fig. 9.13** Effect of packetization on 5 VoIP users with access point capacity of 1 Mbps. **a** End-to-end delay. **b** MOS

we cannot use a packet of lower packet payload size. Table 9.1 contains the details of the optimum packet payload sizes for different network scenarios for our simulated network.

## 9.4   Proposed Packetization Algorithm

After carrying our exhaustive simulations, we arrive at the conclusion that packetization scheme helps to improve the voice quality even in varying network conditions. The optimum values of the packet payload sizes have been summarized in Table 9.1. The values clearly indicate that the optimum number of voice samples in

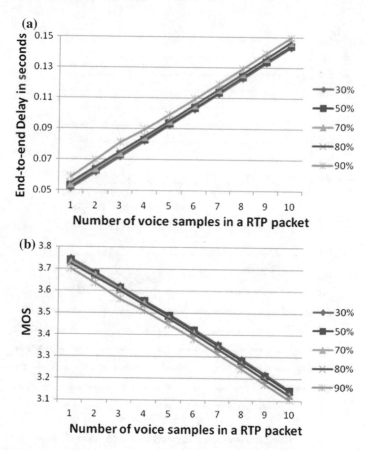

**Fig. 9.14** Effect of packetization on 2 VoIP users with access point capacity of 2 Mbps. **a** End-to-end delay. **b** MOS

the packet payload depends on the capacity of the network, the network congestion as well as the number of VoIP users in the session. For any network connection, the QoS negotiation is done at the setup phase of the connection and the network strives to guarantee this quality excepting severe failures [9]. This suffers from a short-coming because of the changing characteristics of the network. Suppose a high QoS is provisioned at the beginning of the session. However, in the meantime the network condition degrades. At this kind of situation, the QoS is no longer maintained leaving behind unsatisfied users. On the other hand, if a low QoS is guaranteed at the beginning of the session and the network quality enhances in the interim, then the session keeps on providing a poor QoS even though an improvement in the QoS is possible. All these conditions pave way to the requirement of a well-defined algorithm that provides a guideline to vary the number of voice samples in the voice packet payload so as to get the best per-formance out of a VoIP system.

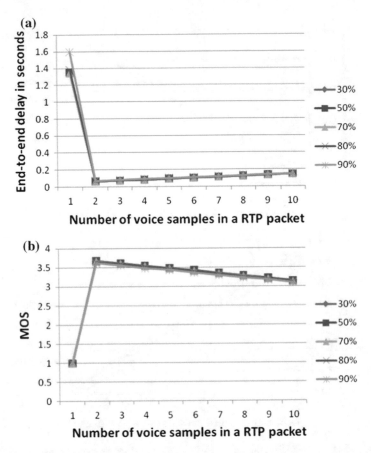

**Fig. 9.15** Effect of packetization on 3 VoIP users with access point capacity of 2s with access point capacity ofMbps. **a** End-to-end delay. **b** MOS

The optimum number of voice samples in the packet payload depends on the capacity of the network, the network congestion as well as the number of VoIP users in the session.

Hence, we propose an algorithm that aims to follow the variations in the network and adjust the number of voice sample frames in the RTP packet payload. Our algorithm can be summarized in the following steps. The steps of the algorithm have also been pictorially represented in Fig. 9.18.

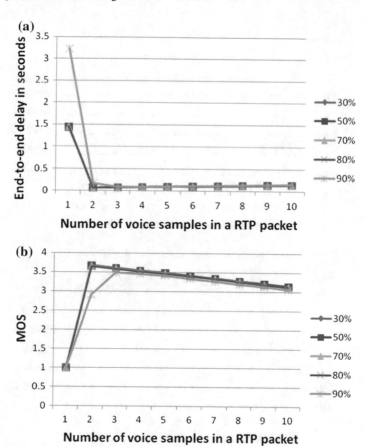

**Fig. 9.16** Effect of packetization on 4 VoIP users with access point capacity of 2 Mbps. **a** End-to-end delay. **b** MOS

Proposed Algorithm

flag = 0;

$n_i$ = number of frames in the $i^{th}$ packet;

$d_i$ = $i^{th}$ packet network delay;     /* $d_i$ is in milliseconds */

If (flag = 0)

{

/* The number of voice sample frames is decreased if the current packet's delay is more than the previous packet's delay or if the network delay is below 100 ms. The second condition aims to minimize oscillations between two values when the delay is not very high. */

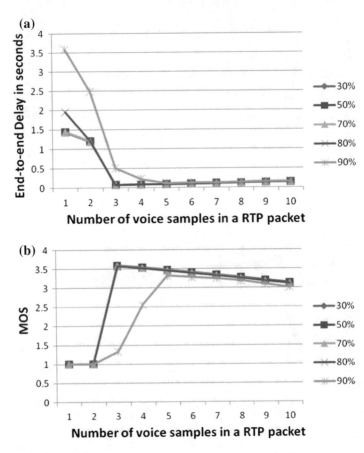

**Fig. 9.17** Effect of packetization on 5 VoIP users with access point capacity of 2 Mbps. **a** End-to-end delay. **b** MOS

**Table 9.1** Optimum number of voice sample frames to be used for different network scenarios

| % of cross traffic | Capacity of the network | | | | | | | |
|---|---|---|---|---|---|---|---|---|
| | 1 Mbps | | | | 2 Mbps | | | |
| | 2 users | 3 users | 4 users | 5 users | 2 users | 3 users | 4 users | 5 users |
| 10 | 1 | 1 | 2 | 3 | 1 | 1 | 1 | 3 |
| 20 | 1 | 2 | 2 | 3 | 1 | 2 | 2 | 3 |
| 30 | 1 | 2 | 2 | 3 | 1 | 2 | 2 | 3 |
| 40 | 1 | 2 | 2 | 3 | 1 | 2 | 2 | 3 |
| 50 | 1 | 2 | 2 | 3 | 1 | 2 | 2 | 3 |
| 60 | 1 | 2 | 3 | 3 | 1 | 2 | 2 | 3 |
| 70 | 1 | 2 | 3 | 3 | 1 | 2 | 2 | 3 |
| 80 | 1 | 2 | 3 | 6 | 1 | 2 | 2 | 3 |
| 90 | 2 | 6 | 9 | N/A | 1 | 2 | 3 | 5 |

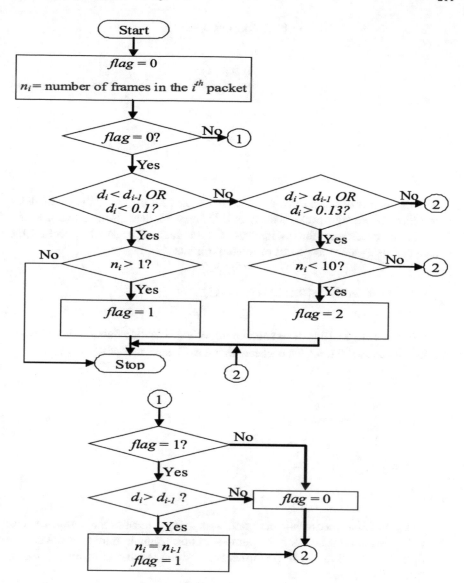

**Fig. 9.18** Flowchart of the proposed algorithm

If ($d_i < d_{i-1}$ OR $d_i < 0.1$)

    {

        If ($n_i > 1$)

        {

            flag = 1;

            $n_i = n_{i-1} - 1$;

        }

    }

/* The number of voice sample frames is increased if the current packet's delay is less than the previous packet's delay or if the network delay is more than 130 ms. The second conditions aim to increase the number of packet size before the end-to-end delay tends to go out of tolerable limits. */

If (($d_i > d_{i-1}$ AND $d_i < 0.1$) OR $d_{i-1} > 0.13$)

    {

If ($n_i < 10$)    /* 10 is the maximum no. of voice sample frames allowed as if the no. is increased further, the packetization delay becomes significantly high */

        {

            flag = 2;

            $n_i = n_{i-1} + 1$;

        }

    }

    }

/* The condition checks whether there is an improvement in the performance of the VoIP session on decreasing the number of voice sample frames. If the change has further degraded the VoIP performance, then the number of sample frames is reverted to the previous value.*/

Else if (flag = 1)

{

If ($d_i > d_{i-1}$ )

{

$n_i = n_{i-1}$;

flag = 2;

}

Else

flag = 0;

}

Else

flag = 0;

/* The flag values of the algorithm are reset, and the algorithm continues to operate till the end of the VoIP session. */

## 9.5   Implementation and Results

### 9.5.1   *Implementation*

The algorithm was implemented in OPNET Modeler 14.5.A and thorough simulations have been carried out in order to test for its efficiency. For this purpose, we created a special network node with the help of OPNET. The protocol stack depicted in Fig. 9.19 was executed inside the created network node. The VoIP, UDP, IP, and MAC modules are the standard models provided by the OPNET library. We modified the RTP module to serve the desired operations. For the purpose, we took the support of the RTP and RTCP protocols, which allowed the sender and receiver, estimate the network variations and adjust accordingly. Then, we incorporated the code developed by us in the appropriate location so as to embed the algorithm in the designed network node.

### 9.5.2   *Results*

In the quest of finding the efficiency of the proposed algorithm, several simulations were carried out using several network capacities. Moreover, the network congestion was also changed during the course of the VoIP session to test the

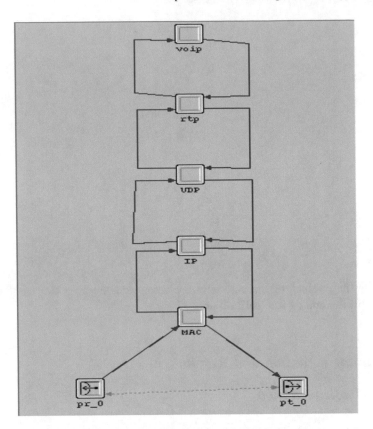

**Fig. 9.19**  Protocol stack of the developed network node in OPNET

adaptability provided by our algorithm. It was found that the algorithm was able to estimate the network traffic variation and select the optimum packet payload size accordingly. The selection of proper packet payload size for various network delays can be seen in Fig. 9.20. Figure 9.21 clearly indicates the reduction in end-to-end

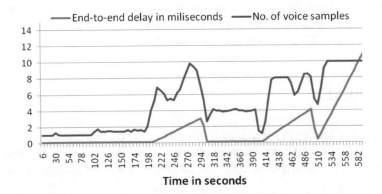

**Fig. 9.20**  Variation in packet payload size w.r.t the variation in end-to-end delay

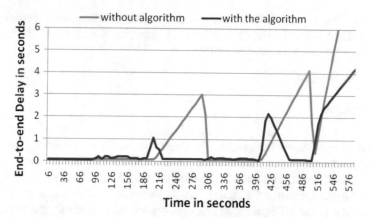

**Fig. 9.21** Improvement achieved by applying the proposed algorithm

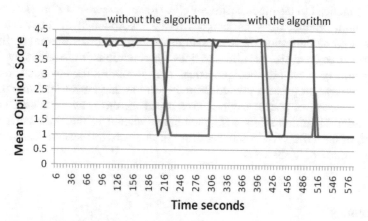

**Fig. 9.22** Improvement in MOS achieved by applying the proposed algorithm

delay attained by applying the algorithm in the VoIP system. It is observed from Figs. 9.20 and 9.21 that after a certain point in network congestion, the maximum allowable packet payload size is reached (at about 520 s). After that point, the end-to-end delay increases with the increase in further network congestion even after application of the algorithm. We have set the limit of 10 to the maximum number of voice samples in a RTP packet payload because 5 ms is required for creating a voice sample. Hence, if the number of voice samples for a single RTP packet is allowed to increase beyond 10, the packetization delay caused during the creation of the packet increases significantly. The condition becomes such that the benefit achieved by the packetization technique in terms of reduction in network delay is nullified by the time consumed in creating the RTP packet. Figure 9.22 can further verify the results, where the comparative results of the MOS values are shown. The results clearly show the enhancement achieved by the application of the algorithm. The algorithm has been tested for various other network congestion scenarios and has provided similar results. Thereby, we have verified the feasibility of the proposed packetization algorithm.

## 9.6   Summary

The work in this chapter has given us an idea of the traffic routed through an already congested network that can be reduced by accommodating more number of voice samples in a RTP packet payload. The method reduces the number of RTP packets required for transmission of a certain message from sender to receiver. Hence, the overhead due to the RTP packet headers gets reduced. As a result, the network can handle and route packets more efficiently and the time consumed by the packet while traveling from the sender to the receiver is also lowered. This reduction in delay is beneficial to the time-sensitive VoIP. However, care must be taken to select the optimum number of voice samples to be incorporated in the voice packet payload because collection of voice samples consumes time. Moreover, the optimum packet payload size for different network conditions is also different. Hence, there should be a well-defined procedure that accomplishes the transition effectively. We have proposed a packetization algorithm that aims to provide the most efficient selection of packet payload size based on the network congestion. The packetization algorithm offers significant improvement in the VoIP performance. The QoS of the VoIP session enhances for varying network scenarios as well as for various number of VoIP users.

## References

1. B. Ngamwongwattana, Effect of packetization on VoIP performance, in *Proceedings of the International Conference on Electrical Engineering/Electronics, Computer, Telecommunications and Information Technology*, pp. 373–376, 14–17 May 2008 (2008)
2. Information Sciences Institute, University of Southern California, Internet Protocol Darpa Internet Program, Protocol Specification, RFC 791, September 1981
3. B.A. Forouzan, S.C. Fegan, *Data communications and Networking* (McGraw-Hill, New York, 2003)
4. H. Schulzrinne, S. Casner, R. Frederick, V. Jacobson. RTP: A Transport Protocol for Real-Time Applications, RFC 3550, July 2003 (2003)
5. Quality of service for voice over IP, Cisco Systems (2001)
6. R.G. Cole, J.H. Rosenbluth, Voice over IP performance monitoring. ACM SIGCOMM Comput. Commun. Rev. **31**(2), 9–24 (2001)
7. The OPNET website. http://www.opnet.com
8. K. Claffy, G.J. Miller, K. Thompson, The nature of the beast: recent traffic measurement from an Internet backbone. In Proceedings of INET'98, Geneva, Switzerland, July 1998 (1998)
9. I. Busse, B. Deffner, H. Schulzrinne, Dynamic QoS control of multimedia applications based on RTP. Comput. Commun. **19**, 49–58 (1996)

# Chapter 10
# VoIP Over Emerging Networks: Case Study with Cognitive Radio Networks

## 10.1 Introduction

Voice over IP (VoIP) technology was established with the integration of Internet and communication technologies in order to reduce the cost of communication and also merge data services with voice. In view of its merits over traditional telephony services, extensive research has been carried out for deploying and maintaining VoIP in practical networks, which has, in turn, lead to a steady rise in the number of VoIP subscribers, and the same has already been observed in detail in the previous chapters. However, as with any technological advancement, VoIP suffers from scalability issues that are aggravated by its stringent Quality of Service (QoS) requirements. Also, with the advent of emerging networks, VoIP may be implemented over different such platforms that include wireless LAN, WiMAX (Worldwide Interoperability for Microwave Access), cellular networks, LTE (Long-Term Evolution) networks, cognitive radio networks (CRN), 5G networks, etc. Each such system has its own sets of standards and regulations and introduces unique challenges that must be addressed while deploying VoIP over such networks. As a case study, this chapter discusses the prospects, challenges, and research perspectives for implementing VoIP services over CRN.

Rapid advancements in the wireless communication sector have led to the advent of emerging smart networks that focus on building an integrated platform for hosting a wide range of applications. With feature richness and user-friendliness as their primary criteria, these bandwidth-savvy applications continue to attract increasing number of subscribers, thereby leading to high data consumption and contributing to the problem of spectrum congestion [1, 2]. This problem severely limits the total system capacity while degrading the application-level QoS, thus posing a serious question to the reliability of such systems. CRN [3–5] promises to solve this spectrum scarcity problem by allowing these applications to access the idle frequency slots in different spectrum bands [6], and in doing so, it increases the overall spectrum utilization.

© Springer International Publishing AG, part of Springer Nature 2019
T. Chakraborty et al., *VoIP Technology: Applications and Challenges*,
Springer Series in Wireless Technology, https://doi.org/10.1007/978-3-319-95594-0_10

At the same time, increasing demand for VoIP services makes it a suitable candidate for CRN technology [7] and is, therefore, the primary focus of research in this chapter. The same is illustrated in Fig. 10.1. An ongoing VoIP communication by SU can be disrupted by the untimely presence of PU in the licensed channel. This implies that the SU must halt VoIP transmission, vacate the current channel, and perform spectrum handoff at the earliest to a suitable idle channel to resume communication. All these operations must be executed in tandem to avoid QoS degradation and subsequent call drop. Therefore, the feasibility study and practical applicability of a real VoIP-based CR system require thorough examination of critical factors and subsequent formulation of design methodologies pertaining to both VoIP and CRN technologies, thus laying the foundation for a new frontier in research in this domain.

> Increasing demand for VoIP services makes it a suitable candidate for CRN technology and is, therefore, the primary focus of research in this chapter.

## 10.2   What Is Cognitive Radio Network?

With a phenomenal increase in the number of wireless subscribers and swift integration of bandwidth-intensive applications with respect to communication, gaming, social networking, and other interactive multimedia services, the problem of spectrum congestion has increased manifold in fixed spectrum assignment policy-based wireless networks [4]. This problem severely limits the overall system capacity and degrades the transmissions for the ongoing users. Consequently, it has led to various studies on spectrum usage across different frequency bands. As recent, Federal Communications Commission (FCC) observations have pointed to large portions of unutilized spectrum in other frequency bands (e.g., television broadcasting bands) [6], the primary focus in this domain is to solve the spectrum scarcity problem of the traditional wireless networks by providing access to these unutilized frequency slots. This has led to the emergence of cognitive radio network

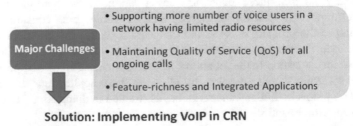

**Fig. 10.1** Packetizer/de-packetizer in a VoIP system

(CRN) [4, 5] that promises to increase the overall spectrum utilization by allowing opportunistic transmissions of different applications in these vacant spectrum bands. The basic idea of a CRN is to allow unlicensed (or secondary) users (SUs) to transmit in the available licensed spectrum bands that are allotted to the licensed (or primary) users (PUs), when the corresponding PUs are absent or idle. The exploiting of spectrum holes by SUs in the absence of PUs is highlighted in Fig. 10.2.

> The problem of spectrum congestion has increased manifold in fixed spectrum assignment policy-based wireless networks and severely limits the overall system capacity and degrades the transmissions for the ongoing users.

The first worldwide standard based on the cognitive radio (CR) technology is the IEEE 802.22 standard [8–10] that mainly focuses on the ultrahigh-frequency (UHF)/very high-frequency (VHF) TV bands between 54 and 862 MHz. This is the licensed frequency spectrum to be used by the SUs on a noninterfering basis, in the absence of PU traffic. The project, formally called the standard for wireless regional area networks (WRANs) uses the fixed point to multipoint-based network topology. Here, a base station (BS) serves as the central managing unit and coordinates with all the subscribed users who are termed as the Consumer Premise Equipments (CPEs). In order to offer protection to the licensed PUs, three mechanisms are suggested that include (i) sensing of PU presence, (ii) database access, and (iii) specially designed beacons. Cognitive capability apart from those included in the SUs is also extended to three significant operations, namely (i) dynamic and adaptive scheduling of quiet periods, (ii) alerting the BS of the presence of PUs by SUs, and (iii) the decision by BS to move one or more subscribers out of their current operating channels.

**Fig. 10.2** Concept of spectrum holes in CRN

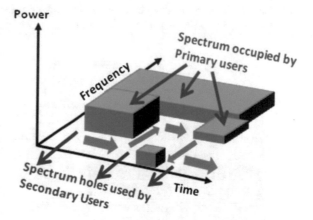

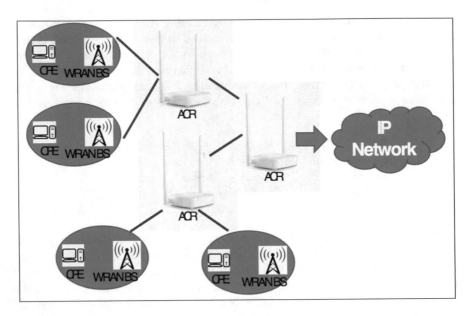

**Fig. 10.3** WRAN architecture

The general architecture of WRAN is shown in Fig. 10.3.

The broad domain of CRN however extends to different licensed bands in addition to the TV bands that are already used in IEEE 802.22 standard. A general classification of the CR systems is illustrated in Fig. 10.4.

Accordingly, spectrum sharing and allocation in CRN can be described under two broad categories. In an underlay-based spectrum sharing scheme [4], SUs spread their transmission over a wide range of frequency bands while maintaining the interference with PUs below the noise floor. On the other hand, the overlay-based spectrum sharing policy [4] supports opportunistic usage of idle channels by SUs in the absence of PUs. Taking into account the absolute priority of a PU over a SU with respect to channel access in overlay-based CRN, a SU must

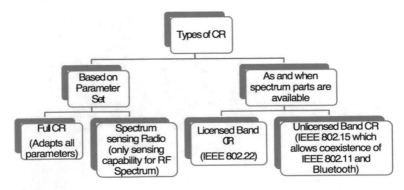

**Fig. 10.4** Types of CR systems

vacate itself on PU arrival and perform spectrum handoff to shift to another idle channel for its transmission. In the absence of a suitable idle channel, this SU is dropped from the network. Thus, arrival of a PU triggers unwanted disruptions in SU transmissions, which may severely degrade the QoS of the underlying applications for SUs. Real-world implementation of CRN, therefore, faces unique challenges in the field of spectrum sensing and analysis, spectrum management, spectrum mobility, and sharing [4], along with the appropriate tuning of wireless device parameters and regulations on spectrum usage. Notwithstanding these complexities, CRN promises to be an effective platform for hosting VoIP applications, specifically when the number of VoIP users is increasing and the integration of VoIP with other services (gaming, marketing, social networking, business, health) is on the rise.

> Real-world implementation of CRN faces unique challenges in the field of spectrum sensing and analysis, spectrum management, spectrum mobility, and sharing, along with the appropriate tuning of wireless device parameters and regulations on spectrum usage.

## 10.3   VoIP Over CRN: Why??

The primary motivation for deploying VoIP services over CRN is driven by two important factors that include firstly, the deterioration in call quality once and after the traditional wireless networks reach their capacity limits with the ever increasing number of VoIP users, and secondly, the opportunistic mode of communication as allowed by CRN through dynamic spectrum management policies. In view of the enormous significance of these technologies in the next-generation wireless communication as highlighted in the previous section, the research community has taken due cognizance of this prospect with several studies being conducted on VoIP-based CR systems, primarily with respect to capacity analysis and traffic modeling. Considering the complexities involved in maintaining strict QoS requirements in such opportunistic networks like CRN, the problem domain as discussed in this chapter has emerged as a promising yet challenging area of the present-day research. Although both CRN and VoIP technologies have emerged as potential winners in their respective fields and attracted widespread research activities, interestingly only limited works have conducted joint studies on such VoIP-based CR systems. This section provides a brief overview of these studies and thereby establishes the motivation for conducting research in this domain.

Although both CRN and VoIP technologies have emerged as potential winners in their respective fields and attracted widespread research activities, interestingly only limited works have conducted joint studies on such VoIP-based CR systems.

It is not without reason that the concept of CR was envisioned in the landmark study by Mitola in [3] keeping in mind the prospective growth of mobile multimedia communications. In this paper, he has discussed the advantages of spectrum pooling in solving the spectrum scarcity crisis and providing adequate bandwidth required for different multimedia applications. His work in 1999 invoked another significant study by Haykin in [11] where the author took over from where Mitola had left his preliminary research on CR. This study has focused on the signal processing aspects of CR and introduced interference temperature as a metric for quantifying and managing the interference in CRN. Both these works subsequently triggered a plethora of contemporary research on CR- and CR-based systems in relation to spectrum sensing and analysis [12–14], management [15, 16], and mobility issues [17, 18].

However, it was not until 2009–2010 when the benefits of implementing VoIP applications over CRN were truly realized and studied by Lee and Cho in [7, 19]. In both these works, the reasons for considering VoIP as one of the "candidate applications" in CRN are discussed. This is followed by an extended analysis of the VoIP capacity in CRN where SUs perform VoIP communication in the absence of PUs. The VoIP traffic is modeled as a simple on–off model and matched with the Markov modulated Poisson process (MMPP) model using the Index for Dispersion of Counts (IDC) matching techniques in [19] and further studied in [7] under the conditions of imperfect spectrum sensing by SUs including false alarms and miss-detections.

It was not until 2009–2010 when the benefits of implementing VoIP applications over CRN were truly realized and studied.

As the vision of executing VoIP calls over CRN finally saw the light of the day, subsequent research studies used this platform to evaluate the capacity from different aspects of the system, for example with respect to retransmissions and multi-channel effects in [20], centralized and distributed channel access schemes in [21], medium access control (MAC) protocols in [22, 23] and scheduling algorithms in [24].

On a parallel front, another research group has carried out several studies [25–28] on VoIP transmissions over CRN. In particular, the VoIP traffic is modeled as a simple on–off model and studied under different PU arrival models (on–off and Poisson distribution) in [25]. It is proved that the system performance metrics (in terms of packet dropping and blocking probabilities, and packet delays) are further degraded under the Poisson arriving PUs, as compared to on–off-based PU traffic, due to higher variance in white space duration under the Poisson model. In addition, using the timescale decomposition technique and the continuous and discrete-time Markov chains, an admission control policy with fractional buffering is proposed in [26] with the focus on increasing the VoIP Erlang capacity for the admitted SUs in the network. This work is extended in [27] where a novel joint packet-level and connection-level model is designed for VoIP traffic. Finally, call admission control (CAC) mechanisms for VoIP SUs are devised in [28] where it is proved that the inclusion of primary resource occupancy information can lead to significant reduction in call drop probabilities for the SUs.

More recently, two significant works [29, 30] have been reported in the literature, which provide actual implementation of real-time applications over CRN using practical test beds. Specifically, the RECOG model in [29] configures the centrally managed access points (APs) with several cognitive functionalities, while bestowing the SUs with two transceivers for simultaneous sensing and transmission, and finally implements VoIP and video streaming applications in the system. On the other hand, the soft real-time model in [30] implements only video-streaming applications in CRN under a specific assumption that the SUs are fully aware of the frequency-hopping-based PU traffic.

A closer look into the literature survey in this section reveals the fact that there is a strong demand for comprehensive studies on QoS aware VoIP communication over CRN. This laid the foundation for the proposed research work in this thesis whose relevance has only grown over these years. Overall, the novelty and significance of this study are established in Fig. 10.5 by drawing a comparative analysis of what has already been achieved in the literature and which aspects are still lacking and require further investigations for ensuring successful implementation of VoIP services over CRN.

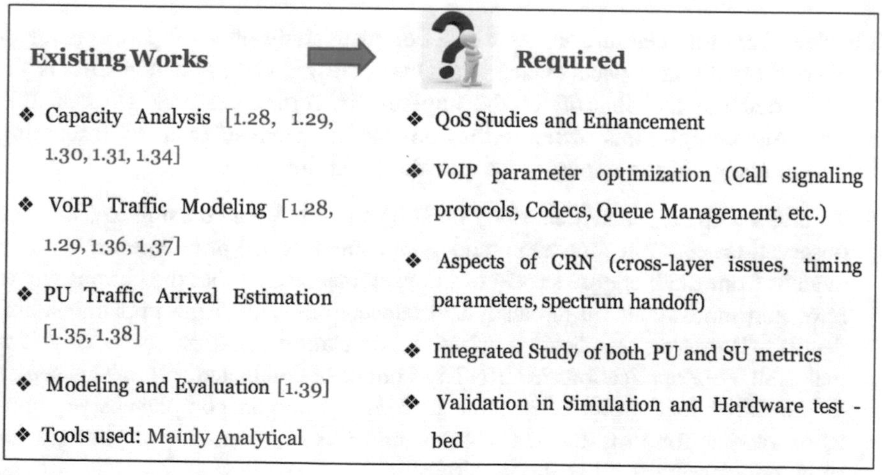

**Existing Works** ➡️ ❓ **Required**

❖ Capacity Analysis [1.28, 1.29, 1.30, 1.31, 1.34]

❖ VoIP Traffic Modeling [1.28, 1.29, 1.36, 1.37]

❖ PU Traffic Arrival Estimation [1.35, 1.38]

❖ Modeling and Evaluation [1.39]

❖ Tools used: Mainly Analytical

❖ QoS Studies and Enhancement

❖ VoIP parameter optimization (Call signaling protocols, Codecs, Queue Management, etc.)

❖ Aspects of CRN (cross-layer issues, timing parameters, spectrum handoff)

❖ Integrated Study of both PU and SU metrics

❖ Validation in Simulation and Hardware test - bed

**Fig. 10.5** Relevance of the proposed research

As the vision of executing VoIP calls over CRN finally saw the light of the day, subsequent research studies used this platform to evaluate the capacity from different aspects of the system.

## 10.4   Challenges Toward Deploying VoIP Over CRN

Thus, it is inferred from the previous section that the integration of VoIP and CRN technologies can prove to be a determining factor toward providing "anytime anywhere" communication to people across the globe, albeit at lesser costs and under lower maintenance overheads. In this regard, the intrinsic design challenges need to be suitably addressed with respect to maintaining an optimal trade-off between PU protection and QoS guaranteed VoIP communication by SUs. Specifically, the objective is to focus on three primary aspects that are highlighted as follows.

- Analytical- and simulation-based studies of the spectrum management policies with respect to sensing, transmission, analysis, decision and mobility and also their suitability toward providing QoS guaranteed VoIP communication.
- Studies involving the design formulation of cross-layer-based adaptive strategies involving VoIP and CRN parameters toward enhancing the overall call quality.
- Practical test bed-based feasibility studies of hosting VoIP applications over CRN without compromising with the call quality for the SUs and without any harmful interference to the PUs.

## 10.5   Key Focus Areas

It is clear from the literature survey that a complete study of VoIP performance in CRN covering all the aspects related to analysis, design, and implementation is yet to be carried out and this drives the impetus for further research work in this domain. Accordingly, this section brings out the key areas of focus by discussing the shortcomings of the relevant works in the literature.

- To begin with, any analytical study must be suitably verified using real life-like observations either in simulation models or using test-bed prototypes. It is pretty evident from the literature survey of the previous section that the current works have laid more focus on the analytical aspects rather than the implementation details. However, application-oriented simulation studies, especially for real-time VoIP applications in CRN have not been conducted so far. This serves as the initial focal point of research toward developing comprehensive simulation models for VoIP-based CRN that will serve as a dedicated platform for conducting further studies in this discipline.

- Focusing on the VoIP parameters, codecs play a significant role toward shaping the VoIP traffic, eventually affecting the VoIP call quality. However, codec switch based on reactive strategy often leads to considerable degradation in user throughput which can be mitigated using proactive approach. Most importantly, none of these works have been executed with respect to CRN. Therefore, the next focus of research is on developing proactive codec adaptation algorithm in order to minimize this throughput degradation and subsequently applying it to VoIP sessions over CRN.

- Next, the CR cycle timing parameters comprising of the sensing interval and transmission duration are considered because suitable configuration of these intervals is integral toward ensuring the success of any CR system. Therefore, after developing and analyzing the simulation models for VoIP users in CRN, the next step is to determine the optimal sensing and transmission durations for protecting the QoS of VoIP SUs, and also configure the transmission time before initiating any VoIP call in a particular channel.

- The ultimate goal of any overlay-based CRN system is to maximize spectrum utilization by allowing SUs to access available spectrum when the PUs are absent. However, maximum system capacity in CRN is limited due to several factors as derived in [31]. It is, thus, implied that the existing policies for the basic CRN fail to record significant increase in performance efficiency once and after the system reaches its maximum capacity limit. This also restricts the system heterogeneity in terms of the different types of users as admitted in CRN from time to time. However, joint studies of system capacity and heterogeneity are yet to be performed specially for VoIP applications in CRN. Thus, the next focus of work is to increase both the system capacity and heterogeneity in terms of total number of users admitted.

- Thereafter, focus shifts from the CR cycle parameters to the studies involving QoS aware spectrum management policies in CRN, e.g., PU-based channel reservation policy. Spectrum mobility is another integral aspect of CRN that requires special attention for real-time applications in view of the disruptions caused during channel switch.

- Finally, the research work must deal with the practical aspects of real-life-like implementation of VoIP-based CR system in the test bed. It is highly imperative that the actual significance of all the existing works in the literature with respect to VoIP-based CRN can only be realized through prototype modeling, followed by test-bed implementation.

## 10.6   System Parameters in CRN

### 10.6.1   Spectral Efficiency

Since the licensed frequency spectrum allocated to the SUs is limited, it has to be utilized efficiently. A given bandwidth is used effectively only when maximum information can be transmitted over it. The term "Spectral Efficiency" is used to describe the rate of information being transmitted over a given bandwidth in specific communication systems. It can be calculated by dividing the total amount of data bits transmitted by the available bandwidth of the underlying channel. If a communication system uses 1 kHz of bandwidth to transmit 1,000 bits per second, then it has a spectral efficiency or bandwidth efficiency of 1 (bit/s)/Hz.

Spectral efficiency assumes special significance for CRN whose primary objective is to increase the overall spectrum utilization by allowing access to the idle spectrum bands. However, in an overlay CRN, there is a trade-off between the level of protection offered to the PUs and the spectral efficiency achieved through incorporation of the SUs. Thus, the system capacity becomes another important issue in the context of spectral efficiency and is discussed next.

> In an overlay CRN, there is a trade-off between the level of protection offered to the PUs and the spectral efficiency achieved through incorporation of the SUs.

### 10.6.2   System Capacity

This is an important metric that indirectly implies the overall spectral efficiency of the CRN and can be defined as the maximum number of simultaneous SUs that can be admitted in a fixed frequency band. Unlike traditional networks, system capacity in CRN can vary depending on several factors including PU arrival rate, PU traffic distribution, variable channel conditions and heterogeneous application requirements of the SUs. In this regard, CRN must incorporate two important aspects; (i) call admission control (CAC) policies and (ii) queuing models. Some works [31] have also expressed the system capacity of CRN in the form of the SU Sum Goodput that is defined as the total amount of SU data successfully transmitted. It is to be noted that while increasing the system capacity can improve the spectral efficiency, it may degrade the energy efficiency of the system which is another important parameter and is described as follows.

> While increasing the system capacity can improve the overall spectral efficiency, it may degrade the energy efficiency of the system.

### 10.6.3   Energy Efficiency

Energy efficiency metric is an important one to be considered during network planning and is defined as the information bits transmitted per unit of transmitted energy. Considering the power constraints of mobile phones which are the primary users of VoIP applications, building energy efficient CRN [32, 33] is highly critical toward ensuring widespread acceptability. In such opportunistic communication models, energy can be lost while sensing as well as during transmission specifically in an overlay CRN. Accordingly, the CR timing cycle comprising of the sensing duration and the transmission duration plays a crucial role toward ensuring "green" communications and is discussed as follows.

> In opportunistic communication models, energy can be lost while sensing as well as during transmission specifically in an overlay CRN.

### 10.6.4   CR Timing Cycle

Every SU in a CRN is required to follow the CR timing cycle that comprises of sensing duration followed by the transmission interval. PU detection is performed in the sensing slot, and if the channel is detected idle, transmission is resumed in the transmission duration. Obviously, there is a trade-off regarding optimal selection of these slots that is explained as follows. Increasing the sensing duration will provide higher protection to PU traffic as the probability of detecting PU presence will also increase. However, this will reduce the transmission time for the SU, thus resulting in loss of spectrum utilization [34]. On the other hand, increasing the transmission slot will ensure QoS guarantees for the VoIP calls by the SUs but will aggravate the risk of interference with the PU traffic. In general, the energy consumed as well as the throughput recorded through successful transmissions must be optimally maintained through proper configuration of these timing intervals.

> In general, the energy consumed as well as the throughput recorded through successful transmissions must be optimally maintained through proper configuration of these timing intervals.

## 10.6.5   Spectrum Handoff Delay

Spectrum mobility is a critical aspect of CRN that determines the long-term sustainability of the SU-based applications in the system. One metric that quantifies this aspect is the spectrum handoff delay [4]. This delay is unique to the CRN and is analogous to the handoff delay as observed in traditional wireless networks. Here, whenever PU arrives in the channel currently occupied by SU, the SU has to quit its transmission for the moment, select an idle channel, and resume transmission in the newly selected channel after performing appropriate channel switching operations. This process is termed as spectrum handoff and the delay incurred therein is the spectrum handoff delay.

The spectrum handoff delay is the variable component delay and therefore adds to the jitter when dealing with real-time packet-switched traffic such as VoIP. It includes several delay components that are described as follows.

  i. Channel switching delay: Actual switching of the channel requires shifting the operational frequency to that of the new channel and incurs a delay overhead on the part of the transceiver carrying out this task. This delay can be reduced through efficient hardware design of the radio terminal.
 ii. Channel consensus delay: This delay follows the channel switching delay and is the time spent for achieving consensus in the new channel among the different SUs who are involved in communication. This consensus may be achieved through message passing and/or with the intervention of the SC node, and thus consumes a noticeable time overhead.
iii. Channel selection delay: Selecting an idle channel during handoff operation depends on the efficiency of the spectrum handoff policy adopted. The more the time spent in selecting this target channel, the higher the disruption in communication by the SU. This time spent is the channel selection delay. Reduction of this delay has been the primary focus of research in this domain of spectrum mobility. This is because this delay not only degrades the QoS for the underlying applications but also increases the probability of dropping from the channel (when the delay exceeds the threshold limit) which is not acceptable.

> Spectrum mobility is a critical aspect of CRN that determines the long-term sustainability of the SU-based applications in the system.

## 10.7   Design Principle: An Overview

An overview of the network infrastructure for VoIP in CRN is provided in Fig. 10.6. Here the VoIP calls are initiated and managed by the unlicensed SUs in the CRN.

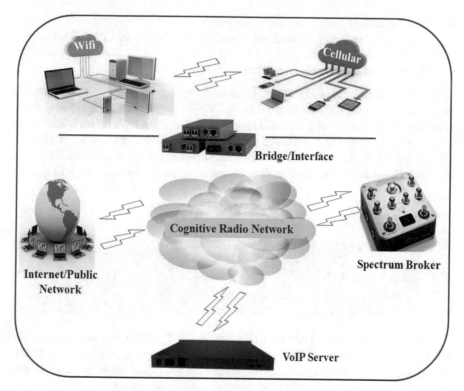

**Fig. 10.6** Network architecture depicting deployment of VoIP services over CRN

The key operations are described as follows.

- SUs request for and obtain default channels (frequency bands) for initiating transmissions in the absence of the licensed PUs.
- Two SUs initiate VoIP communication in the default channel following standard call signaling protocols (SIP or H.323) and start exchanging real-time data in the form of real-time transport protocol (RTP)/real-time transport control protocol (RTCP) packets.
- In between such exchanges, the SUs periodically sense the channel for any PU activity. Three possible cases may occur based on the sensing outcome.

    - Firstly, if the channel is sensed idle, the SUs can resume their ongoing VoIP session and carry on with the communication.
    - Secondly, if the channel is sensed busy, the SUs stop their transmission and perform "spectrum handoff" to another available idle channel where they resume the disrupted VoIP call.
    - Finally, in the third case, if the channel is sensed busy and no other idle channel is available, the VoIP call is dropped by the SUs.

- After the call is discontinued by the end user, the termination of VoIP calls follows the standard procedure as discussed previously in Chap. 2 where the underlying call signaling protocol-based messages are exchanged between the SU caller and callee and the session is finally closed.

The driving factor behind implementation of VoIP over CRN stems from the fact that link utilization, and hence, capacity of VoIP system must be increased along with overall enhancement in the call quality [3]. Let $L$ be the link utilization with respect to time and $M$ be the derived expression from Mean Opinion Score (MOS). Let $f(L, M)$ be the function of $L$ and $M$. Therefore, the objective is to maximize $f(L, M)$. Hence, the objective function can be mathematically denoted by (10.1).

$$T = \max[f(L, M)] \qquad (10.1)$$

The driving factor behind implementation of VoIP over CRN stems from the fact that link utilization and hence, capacity of VoIP system must be increased along with overall enhancement in the call quality.

Active networks like VoIP transmission in CRN embed computational capabilities into conventional networks, thereby massively increasing the complexity. Therefore, simulation has scored over traditional analytical methods in analyzing such active networks [26]. This necessitates appropriate design of models in the simulation platforms based on an underlying principle. In this regard, the basic principle behind proposed simulation model is shown in Fig. 10.7.

PU generates traffic at uniform distribution interval. The SU is the VoIP user and implements G.711 codec [27]. The PU is the licensed user with priority to use the channel. But it does not always occupy channels and SUs as CR users are permitted to use these channels in the absence of PU. An SU senses the assigned channel in the sensing period and starts transmission in the transmission period only when the PU is inactive.

## 10.8   Model Overview for VoIP Deployment Over CRN

In the absence of any established standard for VoIP implementation over CRN, research in this realm is primarily based on simulation studies and mathematical analysis. However, model description of VoIP over CRN in the literature lacks transparency and clarity and also requires proper documentation that must be equipped with suitable verification and validation. It has generally been observed that simulation studies in telecommunication and networking often fall short of credibility due to inappropriate use of pseudorandom generators and lack of an in-depth analysis after conducting the simulations [1]. Therefore, the design of a

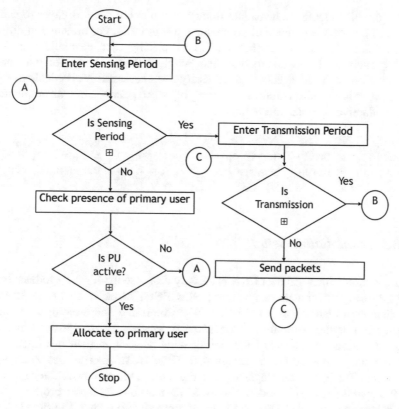

**Fig. 10.7** Flowchart depicting the proposed CR approach where SUs sense of channel for PU presence before transmission

proper simulation model is as crucial as the analysis and optimization of challenges involved in providing QoS aware VoIP transmissions over CRN. Additionally, the simulation output from the designed model must be analyzed using mathematical tools and validated accordingly so as to facilitate their utilization in future research activities globally [2].

The design of a proper simulation model is as crucial as the analysis and optimization of challenges involved in providing QoS aware VoIP transmissions over CRN.

This chapter therefore aims to develop models of VoIP in CRN in different simulation platforms to help researchers with further exploration in this realm. Keeping pace with rapid and continuous development in this domain, several modifications are to be made in the proposed models with the focus on applying novel, adaptive, and cross-layer strategies. Moreover, these models should be modified with suitable mathematical framework in every network layer to support advanced opportunistic mode of communication in CRN. Performance in different types of networks can be further studied by placing such customized nodes in already established network models like WSN, WBAN. On top of that, proper configuration of VoIP agents is a necessity for successful VoIP deployment in CRN, and this includes careful monitoring of codec parameters with implementation of adaptive optimization policies.

This chapter therefore aims to develop models of VoIP in CRN in different simulation platforms to help researchers with further exploration in this realm.

### 10.8.1   Simulation Setup

A simple model for VoIP over CRN is initially designed in OPNET Modeler 16.0.A [28] following distributed architecture [29]. OPNET Modeler 16.0.A is chosen as the simulation platform as it offers highly customizable nodes along with options ranging from traffic distribution and network parameters to cross-layer architecture-based operational modes and collection of wide range of statistical results.

The node model for SU as designed in OPNET Modeler 16.0.A is shown in Fig. 10.8a. The application layer node is the VoIP node responsible for managing VoIP transmissions. This node is followed by real-time transport protocol (RTP), user datagram protocol (UDP), and internet protocol (IP) nodes. The functionalities of network layers are incorporated in process models corresponding to each node in the node model. VoIP_sink node acts as sink for packets already received and processed accordingly. The MAC_Controller node acts as link layer node and cooperates with physical layer node which is involved in sensing, transmission, and reception. Spectrum management functionalities like spectrum sensing and spectrum handoff must work in collaboration with the communication protocols [29], and hence, such cross-layer architecture is implemented in this chapter.

The process model corresponding to MAC_Controller node is highlighted in the Attributes Tab of Fig. 10.8a. It consists of sense and transmits processes that respectively sense and transmit packets according to the design principle as stated in Sect. 3.2. Single-radio architecture is implemented in the sensing principle, where a specific time slot is allocated for spectrum sensing. Thus, only certain accuracy can be guaranteed for spectrum sensing results. It also results in a decrease in spectrum efficiency as some portion of available time slot is used for sensing instead of data transmission [30].

However, the advantage of single-radio architecture is its simplicity and lower cost [31], both of which are essential for low-cost VoIP communication. Moreover, sensing PU presence is based on energy detection-based radiometry [32] or periodogram. The advantages include low computational and implementation

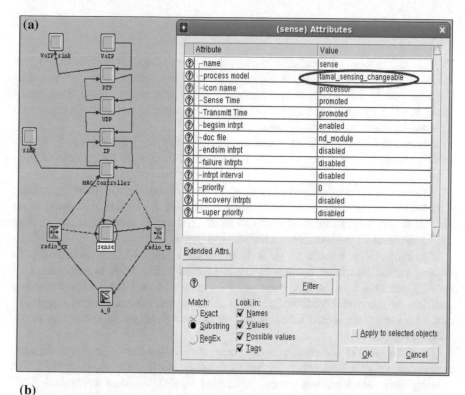

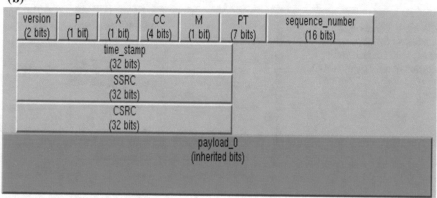

**Fig. 10.8   a** Node model of VoIP over CRN for single channel scenario, and **b** RTP packet format in OPNET modeler 16.0.A

complexities that make the model simpler with less algorithmic delays. It is also generic as receivers need no knowledge on PU's signal [31]. However, it may lead to false alarms [33] and precautions must be taken to eliminate them. Finally, VoIP packets are designed following the protocol formats in each layer. For example, the designed RTP packet is illustrated in Fig. 10.8b.

## 10.8.2    Discussion of Simulation Results

The simulation model is analyzed to evaluate the essential QoS metrics that define the overall call quality in VoIP communication. Many simulation runs are conducted to study the impact of the basic CR cycle parameters (sense time, transmission time) during the ongoing VoIP sessions. Accordingly, the sensing and transmission time for SUs are varied.

As observed from Fig. 10.9, mean end-to-end delay for VoIP calls in SUs rises with increase in sensing period due to increase in waiting time for packets waiting to be transmitted.

This delay decreases with rise in secondary transmission period, thereby creating a favorable environment for VoIP transmission. It is further witnessed from Fig. 10.10 that higher sensing durations increase SU packet loss which further rises with a decrease in the secondary transmission period.

Throughput degradation for SUs is recorded in Fig. 10.11 with increase in sensing period. A sharp decline in received total traffic with higher values of sensing intervals denotes that there must be a maximum bound on the sensing duration for successful VoIP transmission. Finally, it is observed from Fig. 10.12 that jitter rises with increase in sensing period and decrease in secondary transmission period.

Thus, the study of simulation results makes it clear that sensing and transmission intervals have profound effect on the QoS of the VoIP calls. It is evident that sensing time must have both upper and lower bounds to allow successful VoIP

**Fig. 10.9** Variation in end-to-end delay (sample mean) with sensing and transmission intervals of SUs

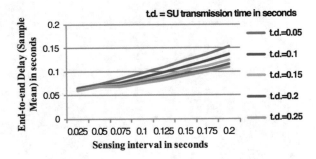

**Fig. 10.10** Variation in packet loss (maximum value) with sensing and transmission intervals of SUs

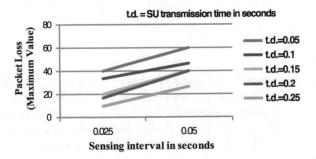

**Fig. 10.11** Variation in traffic received with sensing and transmission of SUs

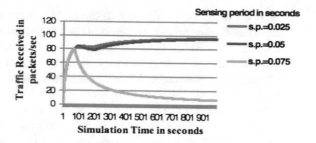

**Fig. 10.12** Variation in jitter (sample mean) with sensing and transmission intervals of SUs

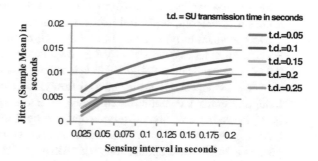

communication without interfering with PU traffic. A solution to this problem includes increasing the sensing period resulting in low SU throughput. Another alternative solution is to use short sensing and transmission cycles which further increase jitter in voice traffic. Therefore, proper configuration of sensing and transmission durations is a tricky problem with respect to the SUs, especially when dealing with the real-time QoS requirements of VoIP traffic.

> Proper configuration of sensing and transmission durations is a tricky problem with respect to the SUs, especially when dealing with the real-time QoS requirements of VoIP traffic.

## 10.9   Applications

Once the CRN is successfully poised to host VoIP services, this will usher in a new era of communication mode having enormous significance in the context of several systems, some of which are discussed as follows.

1. Disaster management systems: Whenever disaster in the form of natural calamities or induced emergencies occurs, communication assumes the foremost priority that will initiate recovery and aid measures. However, in most cases, this communication is severely disrupted in the traditional networks either due to

physical (mechanical) damage to the nodal elements, or due to massive traffic load (enormous number of panic calls) on the network elements. VoIP applications over CRN can take advantage of the existing infrastructure and perform a quick setup of VoIP calls for rapid restoration of communication.

2. Cellular Communication Models: The VoIP-based SU terminal can be implemented in the modern cellular phones after suitably configuring their MAC and PHY layers. This will enable these phones to act as SUs and perform VoIP-based communication in the idle frequency bands of other service providers, when their underlying cellular network becomes congested. The service providers too can use this platform to provide their unused spectrum to the CR users on a temporal basis and at a nominal price. Thus, this technology will prove beneficial to both the end users as well as the service providers.

3. Service Provider Scenarios: As VoIP continues to contribute immensely to the overall communication traffic, there is a serious economic issue as faced by the service providers and is explained as follows. It is already ruled in countries like India that under the conditions of Internet neutrality, any VoIP application can be registered as an IP-based service and can be used by the customers without incurring extra costs. Now, a scenario can be considered where a consumer installs a VoIP-based application (e.g., Skype or WhatsApp) in the IP-enabled mobile phone and uses that application to make voice calls instead of the cellular communication. This is both lucrative and cost-effective for the end user as he has to pay only for the minimal data charges, instead of the cellular call charge which is usually higher, more so for international calls. On the other hand, this implies huge financial loss for the service provider whose main source of revenue is from the cellular communication. Therefore, one way to recover this loss is to lease a portion of the licensed spectrum using the dynamic spectrum leasing (DSL) concept of CRN and allow external users (registered to other domains) to opportunistically access this licensed spectrum for VoIP and other applications.

4. Multiple Device Compatibility Scenarios: Another preferable use is to enable VoIP communication across different device categories including mobile phones, Wi-Fi-enabled laptops, desktop computers and IP-based portable devices. These devices may operate under different domains but can be integrated into using the CR platform for facilitating communication across different categories of users. Also, another application of this technology is the implementation of cost-effective solutions for educational and nonprofit organizations by incorporating low-cost VoIP calls in licensed frequency bands after exploiting the concept of CR Networks.

5. Upcoming 5G Networks: Exponential growth of wireless traffic has led to the formation of the 5G network standard that is promised to satisfy multiple objectives in terms of data rate, latency, cost, energy and spectral efficiency, number of connected devices. Implementing CRN is highly relevant in such scenarios, where novel concepts such as licensed spectrum sharing and dynamic spectrum leasing can be put to good use toward enabling seamless VoIP experience across diverse network categories.

## 10.10 Summary

Evolution in the field of wireless communication has witnessed consistently increasing number of users and wider bandwidth requirement of data and multi-media transmitting technologies that have constantly reduced the availability of frequency spectrum. CRN addresses this problem of spectral congestion by intro-ducing opportunistic usage of the frequency bands that are not heavily occupied by licensed users. In this network, spectrum sensing is done to locate unused spectrum segments and optimally use these segments without harmful interference to the licensed user. Implementation of this technology, therefore, faces unique challenges starting from the capabilities of cognitive radio techniques and the communication protocols that need to be developed for efficient communication to novel spectrum management functionalities such as spectrum sensing, spectrum analysis, spectrum decision, as well as spectrum mobility.

Although considerable progress has been made in the research domain of CRN and its related issues, deployment of real-time applications over CRN has received relatively lesser attention. To be specific, a comprehensive study of VoIP QoS parameters over CRN has not been made till date. Neither do we have a specific CRN management policy for implementing VoIP efficiently over it. This triggers the necessity for active research in this domain in order to exploit the advantages of CRN for the betterment of VoIP and other related real-time transmissions. This also requires modifications in the VoIP domain for its adaptation to the current CRN scenario which varies from other networks in a broad way.

## References

1. Frequency Spectrum Congestion, Available at: http://www.its.bldrdoc.gov/fs-1037/dir-016/_2390.htm (1996)
2. Spectrum Monitoring and Compliance, ICT Regulation Toolkit (2016)
3. J. Mitola, Cognitive radio for flexible mobile multimedia communications, in *Proceedings of IEEE International Workshop on Mobile Multimedia Communications (MoMuC '99)*, San Diego, CA, pp. 3–10 (1999)
4. F. Akyildiz, W.Y. Lee, M.C. Vuran, S. Mohanty, NeXt generation/dynamic spectrum access/ cognitive radio wireless networks: a survey. Comput. Netw. J. **50**, 2127–2159 (2006)
5. Notice of proposed rule making and order: facilitating opportunities for flexible, efficient, and reliable spectrum use employing cognitive radio technologies. Federal Communications Commission (FCC), ET Docket No. 03-108, Feb 2005
6. Spectrum Policy Task Force, Federal Communications Commission (FCC). ET Docket No. 02-135, Nov 2005
7. H. Lee, D. Cho, Capacity improvement and analysis of VoIP service in a cognitive radio system. IEEE Trans. Veh. Technol. **59**(4), 1646–1651 (2010)
8. A.N. Mody, G. Chouinard, IEEE 802.22 wireless regional area networks enabling rural broadband wireless access using cognitive radio technology. IEEE 802.22-10/0073r03, June 2010

9. T. Feng, C. Shilun, Y. Zhen, Wireless regional area networks and cognitive radio. Article No. 3, ZTE Commun (2006)
10. K. Bian, J. Park, Addressing the hidden incumbent problem in 802.22 networks, in *Proceedings of the SDR'09 Technical Conference*, Washington, USA, 1–4 Dec 2009
11. S. Haykin, Cognitive radio: brain-empowered wireless communications. IEEE J. Sel. Areas Commun. **23**(2), 201–220 (2005)
12. A. Ghasemi, E.S. Sousa, Spectrum sensing in cognitive radio networks: requirements, challenges and design trade-offs. IEEE Commun. Mag. **46**(4), 32–39 (2008)
13. D. Cabric, S.M. Mishra, R.W. Brodersen, Implementation issues in spectrum sensing for cognitive radios, in *Conference Record of the Thirty-Eighth Asilomar Conference on Signals, Systems and Computers*, vol. 1, pp. 772–776 (2004)
14. T. Yucek, H. Arslan, A survey of spectrum sensing algorithms for cognitive radio applications. IEEE Commun. Surveys Tutor. **11**(1), 116–130 (2009)
15. I.F. Akyildiz, Spectrum management in cognitive radio networks, in *Proceedings of IEEE International Conference on Wireless and Mobile Computing, Networking and Communications*, pp. xxviii–xxviii, Avignon (2008)
16. I.F. Akyildiz, W.Y. Lee, M.C. Vuran, S. Mohanty, A survey on spectrum management in cognitive radio networks. IEEE Commun. Mag. **46**(4), 40–48 (2008)
17. I. Christian, S. Moh, I. Chung, J. Lee, Spectrum mobility in cognitive radio networks. IEEE Commun. Mag. **50**(6), 114–121 (2012)
18. T. Guo, K. Moessner, Optimal strategy for QoS provision under spectrum mobility in cognitive radio networks, in *Proceedings of IEEE Vehicular Technology Conference (VTC Fall)*, Quebec City, QC, pp. 1–5 (2012)
19. H. Lee, D. Cho, VoIP capacity analysis in cognitive radio system. IEEE Commun. Lett. **13**(6), 393–395 (2009)
20. L. Jiang, T. Jiang, Z. Wang, Z. He, VoIP capacity analysis in cognitive radio system with single/multiple channels, in *Proceedings of 6th International Conference on Wireless Communications Networking and Mobile Computing* (WiCOM), Chengdu, China, 23–25 Sept 2010, pp. 1–4 (2010)
21. S. Gunawardena, Z. Weihua, Voice capacity of cognitive radio networks for both centralized and distributed channel access control, in *Proceedings of IEEE Global Telecommunications Conference (GLOBECOM 2010)*, Miami, Florida, USA, 6–10 Dec 2010, pp. 1–5 (2010)
22. P. Wang, D. Niyato, H. Jiang, Voice-service capacity analysis for cognitive radio networks. IEEE Trans. Veh. Technol. **59**(4), 1779–1790 (2010)
23. P. Wang, D. Niyato, H. Jiang, Voice service support over cognitive radio networks, in *Proceedings of IEEE International Conference on Communications*, pp. 1–5, Dresden (2009)
24. H.S. Hassanein, G.H. Badawy, T.D. Todd, Secondary user VoIP capacity in opportunistic spectrum access networks with friendly scheduling, in *Proceedings of IEEE Wireless Communications and Networking Conference (WCNC)*, Paris, France, 1–4 Apr 2012, pp. 1760–1765 (2012)
25. S. Lirio Castellanos-Lopez, F.A. Cruz-Perez, M.E. Rivero-Angeles, G. Hernandez-Valdez, Performance comparison of VoIP cognitive radio networks under on/off and poisson primary arrivals, in *Proceedings of IEEE 24th International Symposium on Personal Indoor and Mobile Radio Communications (PIMRC)*, London, UK, 8–11 Sept 2013, pp. 3302–3307 (2013). https://doi.org/10.1109/pimrc.2013.6666717
26. S. Lirio Castellanos-Lopez, F.A. Cruz-Perez, M.E. Rivero-Angeles, G. Hernandez-Valdez, VoIP Erlang capacity in coordinated cognitive radio networks, in *Proceedings of 78th IEEE Vehicular Technology Conference (VTC Fall)*, Las Vegas, USA, 2–5 Sept 2013, pp. 1–6 (2013)
27. S. Lirio Castellanos-Lopez, F.A. Cruz-Perez, M.E. Rivero-Angeles, G. Hernandez-Valdez, Joint connection level and packet level analysis of cognitive radio networks with VoIP traffic. IEEE J. Sel. Areas Commun. **32**(3), 601–614 (2014)

28. S. Lirio Castellanos-Lopez, F.A. Cruz-Perez, M.E. Rivero-Angeles, G. Hernandez-Valdez, Impact of the primary resource occupancy information on the performance of cognitive radio networks with VoIP traffic, in *Proceedings of 7th International ICST Conference on Cognitive Radio Oriented Wireless Networks and Communications (CROWNCOM)*, Stockholm, Sweden, 18–20 June 2012, pp. 338–343 (2012)
29. T. Kefeng, K. Kyungtae, X. Yan, S. Rangarajan, P. Mohapatra, RECOG: a sensing-based cognitive radio system with real-time application support. IEEE J. Sel. Areas Commun. **31** (11), 2504–2516 (2013)
30. A. Jain, V. Sharma, B. Amrutur, Soft real time implementation of a cognitive radio testbed for frequency hopping primary satisfying QoS requirements, in *Proceedings of the Twentieth National Conference on Communications (NCC)*, 28 Feb 2014–2 Mar 2014, pp. 1–6 (2014)
31. S. Srinivasa, S. Jafar, How much spectrum sharing is optimal in cognitive radio networks? IEEE Trans. Wirel. Commun. **7**(10), 4010–4018 (2008)
32. G.Y. Li et al., Energy-efficient wireless communications: tutorial, survey, and open issues. IEEE Wirel. Commun. **18**(6), 28–35 (2011)
33. L. Li, X. Zhou, H. Xu, G.Y. Li, D. Wang, A. Soong, Energy-efficient transmission in cognitive radio networks. In: *Proceedings of 7th IEEE Consumer Communications and Networking Conference*, pp. 1–5, Las Vegas, USA (2010)
34. X. Zhou, Y. Li, Y.H. Kwon, A.C.K. Soong, Detection timing and channel selection for periodic spectrum sensing in cognitive radio, in *Proceedings of IEEE Global Telecommunications Conference (GLOBECOM)*, New Orleans, LA, USA, pp. 1–5 (2008)

Printed in the United States
By Bookmasters